Registration #	Variety	Native Location	Variety	Specimen	Genome
727.151.201.122.0	*september-sunii*	E. Asia (E. China, E. Russia, Siberia)	September Sun	Variegated Vine	Kiwi
727.151.201.122.1	*tri-colored-malei*	E. Asia (E. China, E. Russia, Siberia)	Tri-colored Male	Variegated Vine	Kiwi
727.151.201.122.2	*urazainajai*	E. Asia (E. China, E. Russia, Siberia)	Urazainaja	Variegated Vine	Kiwi
727.151.201.122.3	*viktori*	E. Asia (E. China, E. Russia, Siberia)	Viktor	Variegated Vine	Kiwi

727.151.201.13 **SPECIES 13** *latifolia* Wide Leaf Kiwi
727.151.201.14 **SPECIES 14** *macrosperma* Large Seed Kiwi

Registration #	Variety	Native Location	Variety	Specimen	Genome
727.151.201.140.1	*malei*	Asia (China)	Male	Large Seed	Kiwi
727.151.201.140.2	*femalei*	Asia (China)	Female	Large Seed	Kiwi

727.151.201.15 **SPECIES 15** *melanandra* Dark Flower Kiwi

Registration #	Variety	Native Location	Variety	Specimen	Genome
727.151.201.150.1	*malei*	Asia (China)	Male	Dark Flower	Kiwi
727.151.201.150.2	*femalei*	Asia (China)	Female	Dark Flower	Kiwi

727.151.201.16 **SPECIES 16** *polygama* Silver Vine Kiwi

Registration #	Variety	Native Location	Variety	Specimen	Genome
727.151.201.160.1	*polygama*	E. Asia	Original	Silver Vine	Kiwi
727.151.201.160.2	*alabama-seedlingii*	E. Asia	Alabama Seedling	Silver Vine	Kiwi
727.151.201.160.3	*hot-pepperi*	E. Asia	Hot Pepper	Silver Vine	Kiwi
727.151.201.160.4	*pavel-malei*	E. Asia	Pavel Male	Silver Vine	Kiwi
727.151.201.160.5	*vera's-pridei*	E. Asia	Vera's Pride	Silver Vine	Kiwi

727.151.201.17 **SPECIES 17** *purpurea* Purple Kiwi

Registration #	Variety	Native Location	Variety	Specimen	Genome
727.151.201.170.1	*purpurea*	Asia (China)	Original	Purple	Kiwi
727.151.201.170.2	*hardy-redii*	Asia (China)	Hardy Red	Purple	Kiwi

Attorney Family (*Clusiaceae*)

Family 4 *Clusiaceae* **Attorney**
GENUS *Aristoclesia* (Synonym for *Platonia*)

AF875245

OILNUT (*Calophyllum*)

GENUS 01 *Calophyllum* Oilnut
727.151.401.01 **SPECIES 01** *antillanum* Antilles Oilnut

Registration #	Variety	Native Location	Variety	Specimen	Genome
727.151.401.010.1	*antillanum*	W. Indies (Antilles)	Original	Antilles	Oilnut
727.151.401.010.2	*calabra*	W. Indies (Antilles)	Calabra	Antilles	Oilnut

727.151.401.02 **SPECIES 02** *brasiliense* Brazil Oilnut
727.151.401.03 **SPECIES 03** *calabra* Calabra Oilnut
727.151.401.04 **SPECIES 04** *cholobtaches* Chola Oilnut
727.151.401.05 **SPECIES 05** *inophyllum* Ino Oilnut
727.151.401.06 **SPECIES 06** *lucidum* Lucid Oilnut

MANGOSTEEN (*Garcinia*)

GENUS 02 *Garcinia* Mangosteen
727.151.402.01 **SPECIES 01** *cowa* Cowa Mangosteen
727.151.402.02 **SPECIES 02** *dulcis* Gourka Mangosteen

Registration #	Variety	Native Location	Variety	Specimen	Genome
727.151.402.010.1	*dulcis*	.	Original	Gourka	Mangosteen
727.151.402.010.2	*rata*	.	Rat	Gourka	Mangosteen

727.151.402.03 **SPECIES 03** *gambogia* Yellow Gamboge Mangosteen
727.151.402.04 **SPECIES 04** *hanburyi* Hanbury Mangosteen
727.151.402.05 **SPECIES 05** *hombroniana* Hombron Mangosteen
727.151.402.06 **SPECIES 06** *indica* India Mangosteen
727.151.402.07 **SPECIES 07** *kola* Kola Mangosteen
727.151.402.08 **SPECIES 08** *livingstonei* African Imbe Mangosteen
727.151.402.09 **SPECIES 09** *macrophylla* Charachuala Mangosteen

The Complete Book of Fruits, ISBN:0670-86752-7 states that for *Garcinia mangostana*, "The fruits are round purplish brown, about apple size, with a rosette of dead petals around the stalk and an odd flower shaped button on the other end. If you cut the rind around the fruits circumference, the top can be lifted off to reveal about half

727.151.402.13 **SPECIES 13** *myrtifolia* Myrtle Leaf Mangosteen
727.151.402.14 **SPECIES 14** *xanthochymus* Dark Yellow Gamboge Mangosteen

Registration #	Variety	Native Location	Variety	Specimen	Genome
727.151.402.140.1	*xanthochymus*	S. Asia (N. India, W. Himalaya)	Original	Dark Yellow Gamboge	Mangosteen
727.151.402.140.2	*himalayana*	S. Asia (N. India, W. Himalaya)	Himalayan	Dark Yellow Gamboge	Mangosteen
727.151.402.140.3	*tinctoria*	S. Asia (N. India, W. Himalaya)	Tincture	Dark Yellow Gamboge	Mangosteen

MAMMY APPLE (*Mammea*)

GENUS 03 *Mammea* Mammy Apple
727.151.403.01 **SPECIES 01** *africana* African Mammy Apple
727.151.403.02 **SPECIES 02** *americana* American Mammy Apple

BACURY PEACH (*Platonia*)

GENUS 04 *Platonia* Bacury Peach
727.151.404.01 **SPECIES 01** *esculenta* Garden Bacury Peach

Registration #	Variety	Native Location	Variety	Specimen	Genome
727.151.404.010.1	*esculenta*	.	Original	Garden	Bacury Peach
727.151.404.010.2	*insignis*	.	Insignia	Garden	Bacury Peach

AMERICAN MANGOSTEEN (*Rheedia*)

GENUS 05 *Rheedia* American Mangosteen
727.151.405.01 **SPECIES 01** *aristata* Awn American Mangosteen
727.151.405.02 **SPECIES 02** *edulis* Garden American Mangosteen

Registration #	Variety	Native Location	Variety	Specimen	Genome
727.151.405.020.1	*edulis*	W. Indies	Original	Garden	American Mangosteen
727.151.405.020.2	*lemon-dropii*	W. Indies	Lemon Drop	Garden	American Mangosteen

727.151.405.03 **SPECIES 03** *hessii* Lemon Sap American Mangosteen
727.151.405.04 **SPECIES 04** *lateriflora* Wide Flower American Mangosteen
727.151.405.05 **SPECIES 05** *madruno* Equador American Mangosteen
727.151.405.06 **SPECIES 06** *portoricensis* Puerto Rico American Mangosteen

Registration #	Variety	Native Location	Variety	Specimen	Genome
727.151.405.060.1	*portoricensis*	W. Indies (Puerto Rico)	Original	Puerto Rico	American Mangosteen
727.151.405.060.2	*acuminata*	W. Indies (Puerto Rico)	Pointed Leaf	Puerto Rico	American Mangosteen

Dillen Order (*Dilleniales*)

Order 2 *Dilleniales* Dillen

Dillen Family (*Dilleniaceae*)

Family 1 *Dilleniaceae* Dillen

Flora, ISBN:1-74048-097-X, page 501, states that for *Dillenia* "Fleshy star shaped fruit with 5 or 8 segments".

DILLEN STARFRUIT (*Dillenia*)

GENUS 01 *Dillenia* Dillen Starfruit
Flora, ISBN:1-74048-097-X, page 501, states that for *Dillenia alata* "Red fruit, fleshy part edible.
727.152.101.01 **SPECIES 01** *alata* Queensland Red Beech Dillen Starfruit
727.152.101.02 **SPECIES 02** *indica* Chulta Dillen Starfruit
727.152.101.03 **SPECIES 03** *ovata* Oval Leaf Dillen Starfruit
727.152.101.04 **SPECIES 04** *philippinensis* Philippine Dillen Starfruit
727.152.101.05 **SPECIES 05** *suffruticosa* Shrubby Dillen Starfruit

CASKFRUIT (*Doliocarpus*)

GENUS 02 *Doliocarpus* Caskfruit
727.152.102.01 **SPECIES 01** *clainoides* Clain Caskfruit
727.152.102.02 **SPECIES 02** *major* Garden Caskfruit

Registration #	Variety	Native Location	Variety	Specimen	Genome
727.152.102.020.1	*major*	W. Indies	Original	Garden	Caskfruit
727.152.102.020.2	*brevipedicellatus*	W. Indies	Short Pedicell	Garden	Caskfruit

727.152.102.02 **SPECIES 02** *olivaceus* Green Caskfruit

PINZON CASKFRUIT (*Pinzona*)

GENUS 03 *Pinzona* Pinzon Caskfruit
727.152.103.01 **SPECIES 01** *coriacea* Bejuco De Agua Pinzon Caskfruit

Lecy Order (*Lecythidales*)

Order 3 *Lecthidiales* Lecy

Lecy *Lecythidaceae* is a family of tropical trees found in Africa, Madagascar, Southeast Asia, and C. & S. America. It reaches its greatest diversity in C. & S. America. A comprehensive dataset of the C. & S. America species is found in the Virtual Herbarium of The New York Botanical Garden, www.sweetgum.nybg.org/lp. Botanica ISBN:3-8331-1253-0, page 959, places the Lecy *Lecythidaceae* family under the Tea *Theopsida* Class. The Complete Book of Fruits, ISBN:0670-86752-7, page, 207, places the Lecy *Lecythidaceae* family under the Myrtle *Myrtopsida* Class. This book follows the Botanica (i.e. the Dalgren/Cronquist) Classification of the Tea *Theopsida* Class.

Lecy Family (*Lecythidaceae*)

Family 1 *Lecythidaceae* Lecy

GENUS 01 *Lecythis* Lecy

727.153.101.01 **SPECIES 01** *alutacea* Leather Lecy

Registration #	Variety	Native Location	Variety	Specimen	Genome
727.153.101.010.1	*alutacea*	S. America (E. Venezuela to Surinam & Pará, N. Brazil)	Original	Leather	Lecy
727.153.101.010.2	*karuaiensis*	S. America (E. Venezuela to Surinam & Pará, N. Brazil)	Karuai	Leather	Lecy

727.153.101.02 **SPECIES 01** *ampla* Ample Lecy

Registration #	Variety	Native Location	Variety	Specimen	Genome
727.153.101.020.1	*ampla*	C. America (Nicaragua) to S. America (Ecuador)	Original	Ample	Lecy
727.153.101.020.2	*ampullaria*	C. America (Nicaragua) to S. America (Ecuador)	Ampule	Ample	Lecy
727.153.101.020.3	*armilensis*	C. America (Nicaragua) to S. America (Ecuador)	Armila	Ample	Lecy
727.153.101.020.4	*bogotensis*	C. America (Nicaragua) to S. America (Ecuador)	Bogota	Ample	Lecy
727.153.101.020.5	*boyacensis*	C. America (Nicaragua) to S. America (Ecuador)	Boyaca	Ample	Lecy
727.153.101.020.6	*costaricensis*	C. America (Nicaragua) to S. America (Ecuador)	Costa Rica	Ample	Lecy
727.153.101.020.7	*curranii*	C. America (Nicaragua) to S. America (Ecuador)	Curran	Ample	Lecy

727.153.101.03 **SPECIES 03** *barnebyi* Barneby Lecy

727.153.101.04 **SPECIES 04** *brancoensis* Branco Lecy

727.153.101.05 **SPECIES 05** *chartacea* Paper Lecy

Registration #	Variety	Native Location	Variety	Specimen	Genome
727.153.101.050.1	*chartacea*	S. America (N. Brazil to Colombia)	Original	Paper	Lecy
727.153.101.050.2	*calyce*	S. America (N. Brazil to Colombia)	Calyce	Paper	Lecy
727.153.101.050.3	*cupularis*	S. America (N. Brazil to Colombia)	Cupule	Paper	Lecy
727.153.101.050.4	*laevicula*	S. America (N. Brazil to Colombia)	Smooth	Paper	Lecy
727.153.101.050.5	*langsdorffii*	S. America (N. Brazil to Colombia)	Langsdorff	Paper	Lecy
727.153.101.050.6	*marawynensis*	S. America (N. Brazil to Colombia)	Marawyn	Paper	Lecy
727.153.101.050.7	*monosperma*	S. America (N. Brazil to Colombia)	Single Seed	Paper	Lecy
727.153.101.050.8	*rigida*	S. America (N. Brazil to Colombia)	Rigid	Paper	Lecy
727.153.101.050.9	*rorida*	S. America (N. Brazil to Colombia)	Rorippen	Paper	Lecy
727.153.101.051.0	*spruceana*	S. America (N. Brazil to Colombia)	Spruce	Paper	Lecy

727.153.101.06 **SPECIES 06** *confertiflora* Crowded Flower Lecy

727.153.101.07 **SPECIES 07** *corrugata* Wrinkled Lecy

Registration #	Variety	Native Location	Variety	Specimen	Genome
727.153.101.070.1	*corrugata*	S. America (Tropical)	Original	Wrinkled	Lecy
727.153.101.070.2	*cognata*	S. America (Tropical)	Related	Wrinkled	Lecy
727.153.101.070.3	*martinii*	S. America (Tropical)	Martin	Wrinkled	Lecy
727.153.101.070.4	*rosea*	S. America (Tropical)	Rose	Wrinkled	Lecy
727.153.101.070.5	*rubicunda*	S. America (Tropical)	Rubicund	Wrinkled	Lecy
727.153.101.070.6	*salebrosa*	S. America (Tropical)	Salebrose	Wrinkled	Lecy
727.153.101.070.7	*venusta*	S. America (Tropical)	Vein	Wrinkled	Lecy

727.153.101.08 **SPECIES 08** *gracieana* Gracie Lecy

727.153.101.09 **SPECIES 09** *holcogyne* Holcogyne Lecy

727.153.101.10 **SPECIES 10** *idatimon* Idatimon Lecy

Registration #	Variety	Native Location	Variety	Specimen	Genome
727.153.101.100.1	*idatimon*	S. America (Suriname to Maranhão, Brazil)	Original	Idatimon	Lecy
727.153.101.100.2	*amara*	S. America (Suriname to Maranhão, Brazil)	Bitter	Idatimon	Lecy
727.153.101.100.3	*lutea*	S. America (Suriname to Maranhão, Brazil)	Lute Yellow	Idatimon	Lecy

727.153.101.11 **SPECIES 11** *lanceolata* Lance Leaf Lecy

Registration #	Variety	Native Location	Variety	Specimen	Genome
727.153.101.110.1	*lanceolata*	S. America (E. Brazil)	Original	Lance Leaf	Lecy
727.153.101.110.2	*biserrata*	S. America (E. Brazil)	Biserrate	Lance Leaf	Lecy
727.153.101.110.3	*coxiana*	S. America (E. Brazil)	Cox	Lance Leaf	Lecy

Registration #	Variety	Native Location	Variety	Specimen	Genome
727.153.101.110.4	*lacunosa*	S. America (E. Brazil)	Pitted	Lance Leaf	Lecy
727.153.101.110.5	*limbata*	S. America (E. Brazil)	Limber	Lance Leaf	Lecy
727.153.101.110.6	*madagascariensis*	S. America (E. Brazil)	Madagascar	Lance Leaf	Lecy
727.153.101.110.7	*miersii*	S. America (E. Brazil)	Miers	Lance Leaf	Lecy
727.153.101.110.8	*minor*	S. America (E. Brazil)	Lesser	Lance Leaf	Lecy
727.153.101.110.9	*ollaria*	S. America (E. Brazil)	Olla	Lance Leaf	Lecy
727.153.101.111.0	*platyzone*	S. America (E. Brazil)	Flat Zone	Lance Leaf	Lecy
727.153.101.111.1	*pohlii*	S. America (E. Brazil)	Pohl	Lance Leaf	Lecy
727.153.101.111.2	*tuberculata*	S. America (E. Brazil)	Tubercle	Lance Leaf	Lecy
727.153.101.111.3	*variolata*	S. America (E. Brazil)	Variole	Lance Leaf	Lecy
727.153.101.111.4	*vasiformis*	S. America (E. Brazil)	Vase	Lance Leaf	Lecy
727.153.101.12	**SPECIES 12** *lurida*	Smokey Yellow Lecy			
Registration #	**Variety**	**Native Location**	**Variety**	**Specimen**	**Genome**
727.153.101.120.1	*lurida*	S. America (Piauí to Rio de Janeiro, Brazil)	Original	Smokey Yellow	Lecy
727.153.101.120.2	*jarana*	S. America (Piauí to Rio de Janeiro, Brazil)	Jaran	Smokey Yellow	Lecy
727.153.101.120.3	*latifolia*	S. America (Piauí to Rio de Janeiro, Brazil)	Wide Leaf	Smokey Yellow	Lecy
727.153.101.13	**SPECIES 13** *mesophylla*	Middle Leaf Lecy			
727.153.101.14	**SPECIES 14** *minor*	Lesser Lecy			
Registration #	**Variety**	**Native Location**	**Variety**	**Specimen**	**Genome**
727.153.101.140.1	*minor*	S. America (Venezuela to N. Colombia)	Original	Lesser	Lecy
727.153.101.140.2	*bipartita*	S. America (Venezuela to N. Colombia)	Two Part	Lesser	Lecy
727.153.101.140.3	*elliptica*	S. America (Venezuela to N. Colombia)	Elliptic Leaf	Lesser	Lecy
727.153.101.140.4	*magdalenica*	S. America (Venezuela to N. Colombia)	Magdalene	Lesser	Lecy
727.153.101.140.5	*purdiei*	S. America (Venezuela to N. Colombia)	Purdie	Lesser	Lecy
727.153.101.15	**SPECIES 15** *ollaria*	Olla Lecy			
Registration #	**Variety**	**Native Location**	**Variety**	**Specimen**	**Genome**
727.153.101.150.1	*ollaria*	S. America (N. Venezuela)	Original	Olla	Lecy
727.153.101.150.2	*cordata*	S. America (N. Venezuela)	Heart Leaf	Olla	Lecy
727.153.101.16	**SPECIES 16** *parvifructa*	Small Fruit Lecy			
727.153.101.17	**SPECIES 17** *persistens*	Persistent Lecy			
Registration #	**Variety**	**Native Location**	**Variety**	**Specimen**	**Genome**
727.153.101.170.1	*persistens*	S. America (Guianas to Amapá, Brazil)	Original	Persistent	Lecy
727.153.101.170.2	*alba*	S. America (Guianas to Amapá, Brazil)	White	Persistent	Lecy
727.153.101.170.3	*aurantiaca*	S. America (French Guiana)	Orange	Persistent	Lecy
727.153.101.18	**SPECIES 18** *pisonis*	Pea Fruit Lecy			
Registration #	**Variety**	**Native Location**	**Variety**	**Specimen**	**Genome**
727.153.101.180.1	*pisonis*	S. America (Brazil)	Original	Pea Fruit	Lecy
727.153.101.180.2	*amapaensis*	S. America (Brazil)	Amapá	Pea Fruit	Lecy
727.153.101.180.3	*amazonum*	S. America (Brazil)	Amazon	Pea Fruit	Lecy
727.153.101.180.4	*densa*	S. America (Brazil)	Dense	Pea Fruit	Lecy
727.153.101.180.5	*hoppiana*	S. America (Brazil)	Hopp	Pea Fruit	Lecy
727.153.101.180.6	*marcgraaviana*	S. America (Brazil)	Marcgraav	Pea Fruit	Lecy
727.153.101.180.7	*ollaria*	S. America (Brazil)	Olla	Pea Fruit	Lecy
727.153.101.180.8	*paraënsis*	S. America (Brazil)	Pará	Pea Fruit	Lecy
727.153.101.180.9	*pilaris*	S. America (Brazil)	Soft Hair	Pea Fruit	Lecy
727.153.101.181.0	*setifera*	S. America (Brazil)	Bristle	Pea Fruit	Lecy
727.153.101.181.1	*sphaeroides*	S. America (Brazil)	Round	Pea Fruit	Lecy
727.153.101.181.2	*tenuifolia*	S. America (Brazil)	Tenuous Leaf	Pea Fruit	Lecy
727.153.101.181.3	*urnigera*	S. America (Brazil)	Urn	Pea Fruit	Lecy
727.153.101.181.4	*usitata*	S. America (Brazil)	Useful	Pea Fruit	Lecy
727.153.101.181.5	*velloziana*	S. America (Brazil)	Fleece	Pea Fruit	Lecy
727.153.101.19	**SPECIES 19** *pneumatophora*	Lung Lecy			
727.153.101.20	**SPECIES 20** *poiteaui*	Poiteau Lecy			
Registration #	**Variety**	**Native Location**	**Variety**	**Specimen**	**Genome**
727.153.101.200.1	*poiteaui*	S. America (Brazil)	Original	Poiteau	Lecy
727.153.101.200.2	*racemiflora*	S. America (Brazil)	Raceme	Poiteau	Lecy
727.153.101.21	**SPECIES 21** *prancei*	Prance Lecy			
727.153.101.22	**SPECIES 22** *retusa*	Notched Tip Lecy			
727.153.101.23	**SPECIES 23** *schomburgkii*	Schomburg Lecy			
727.153.101.24	**SPECIES 24** *schwackei*	Schwacke Lecy			
727.153.101.25	**SPECIES 25** *serrata*	Sawtooth Lecy			
727.153.101.26	**SPECIES 26** *tuyrana*	Tuyer Lecy			
Registration #	**Variety**	**Native Location**	**Variety**	**Specimen**	**Genome**

Registration #	Variety	Native Location	Variety	Specimen	Genome
727.153.101.260.1	*tuyrana*	C. America (E. Panama) to S. America (N. Ecuador)	Original	Tuyer	Lecy
727.153.101.260.2	*calycocarpa*	C. America (E. Panama) to S. America (N. Ecuador)	Calico Fruit	Tuyer	Lecy
727.153.101.260.3	*elata*	C. America (E. Panama) to S. America (N. Ecuador)	Elated	Tuyer	Lecy
727.153.101.260.4	*melliana*	C. America (E. Panama) to S. America (N. Ecuador)	Honey	Tuyer	Lecy

The Complete Book of Fruits, ISBN:0670-86752-7, page, 207, states that for *Lecythis zabucajo*; "Many consider the Sapucaya Nut superior" [to the Brazil Nut]

SAPUCAYA NUT (*Lecythis zabucajo*)

727.153.101.27 **SPECIES 27** *zabucajo* Sapucaya Nut Lecy

Registration #	Variety	Native Location	Variety	Specimen	Genome
727.153.101.270.1	*zabucajo*	S. America (Ecuador to French Guiana)	Original	Sapucaya Nut	Lecy
727.153.101.270.2	*crassinoda*	S. America (Ecuador to French Guiana)	Crass Node	Sapucaya Nut	Lecy
727.153.101.270.3	*davisii*	S. America (Ecuador to French Guiana)	Davis	Sapucaya Nut	Lecy
727.153.101.270.4	*gracilipes*	S. America (Ecuador to French Guiana)	Slender	Sapucaya Nut	Lecy
727.153.101.270.5	*hians*	S. America (Ecuador to French Guiana)	Hian	Sapucaya Nut	Lecy
727.153.101.270.6	*lecomtei*	S. America (Ecuador to French Guiana)	Lecomte	Sapucaya Nut	Lecy
727.153.101.270.7	*tumefacta*	S. America (Ecuador to French Guiana)	Swollen	Sapucaya Nut	Lecy
727.153.101.270.8	*validissima*	S. America (Ecuador to French Guiana)	Valid	Sapucaya Nut	Lecy

BRAZIL NUT (*Bertholletia*)

GENUS 02 *Bertholletia* Brazil Nut

727.153.102.01 **SPECIES 01** *excelsa* Excelsior Brazil Nut

Registration #	Variety	Native Location	Variety	Specimen	Genome
727.153.102.010.1	*excelsa*	S. America (Tropical)	Original	Excelsior	Brazil Nut
727.153.102.010.2	*nobilis*	S. America (Tropical)	Noble	Excelsior	Brazil Nut

CANNONBALL APPLE (*Couroupita*)

GENUS 03 *Couroupita* Cannonball Apple

727.153.103.01 **SPECIES 01** *antillana* Antilles Cannonball Apple

Flora, ISBN:1-74048-097-X, page 429, states that "Although these trees are fairly rare in cultivation, one species – the cannonball tree *Couroupita guianensis* – is grown in the USA for its spectacular and remarkable fruits. Brownish ball-like capsules mature to red pulp, they have a disagreeable odor".

727.153.103.02 **SPECIES 02** *guianensis* Guiana Cannonball Apple

Registration #	Variety	Native Location	Variety	Specimen	Genome
727.153.103.020.1	*guianensis*	S. America (Guyana, Suriname, French Guiana)	Original	Guiana	Cannonball Apple
727.153.103.020.2	*acreensis*	S. America (Guyana, Suriname, French Guiana)	Acre	Guiana	Cannonball Apple
727.153.103.020.3	*froesii*	S. America (Guyana, Suriname, French Guiana)	Froes	Guiana	Cannonball Apple
727.153.103.020.4	*idolica*	S. America (Guyana, Suriname, French Guiana)	Idol	Guiana	Cannonball Apple
727.153.103.020.5	*membranacea*	S. America (Guyana, Suriname, French Guiana)	Membrane	Guiana	Cannonball Apple
727.153.103.020.6	*surinamensis*	S. America (Suriname)	Suriname	Guiana	Cannonball Apple

727.153.103.03 **SPECIES 03** *nicaraguarensis* Nicaragua Cannonball Apple

Registration #	Variety	Native Location	Variety	Specimen	Genome
727.153.103.030.1	*nicaraguarensis*	C. America (Nicaragua)	Original	Nicaragua	Cannonball Apple
727.153.103.030.2	*cutteri*	C. America (Nicaragua)	Cutter	Nicaragua	Cannonball Apple
727.153.103.030.3	*darienensis*	C. America (Nicaragua)	Darien	Nicaragua	Cannonball Apple
727.153.103.030.4	*magnifica*	C. America (Nicaragua)	Magnificent	Nicaragua	Cannonball Apple
727.153.103.030.5	*odoratissima*	C. America (Nicaragua)	Scented	Nicaragua	Cannonball Apple
727.153.103.030.6	*parviflora*	C. America (Nicaragua)	Small Flower	Nicaragua	Cannonball Apple

727.153.103.04 **SPECIES 04** *peruviana* Peru Cannonball Apple

727.153.103.05 **SPECIES 05** *saintcroixiana* St. Croix Cannonball Apple

727.153.103.06 **SPECIES 06** *subsessilis* Sessile Cannonball Apple

Registration #	Variety	Native Location	Variety	Specimen	Genome
727.153.103.060.1	*subsessilis*	S. America (N. Brazil, N. Peru)	Original	Sessile	Cannonball Apple
727.153.103.060.2	*amazonica*	S. America (N. Brazil)	Amazon	Sessile	Cannonball Apple
727.153.103.060.3	*elata*	S. America (N. Peru)	Elated	Sessile	Cannonball Apple

727.153.103.07 **SPECIES 07** *venezuelensis* Venezuela Cannonball Apple

COCKY APPLE (*Planchonia*)

GENUS 04 *Planchonia* Cocky Apple

727.153.104.01 **SPECIES 01** *andamanica* Andaman Island Cocky Apple

727.153.104.02 **SPECIES 02** *brevistipitata* Short Stipple Cocky Apple

727.153.104.03 **SPECIES 03** *careya* Carey Cocky Apple

Registration #	Variety	Native Location	Variety	Specimen	Genome
727.153.104.030.1	*careya*	N. Australia to Pac. Is. (New Guinea)	Original	Carey	Cocky Apple

Registration #	Variety	Native Location	Variety	Specimen	Genome
727.153.104.030.2	*crenata*	N. Australia to Pac. Is. (New Guinea)	Scalloped	Carey	Cocky Apple

727.153.104.04 **SPECIES 04** *grandis* West Malaysia Cocky Apple
727.153.104.05 **SPECIES 05** *papuana* Papua New Guinea Cocky Apple
727.153.104.06 **SPECIES 06** *spectabilis* Philippine Carey Cocky Apple
727.153.104.07 **SPECIES 07** *sundaica* Greater Sunda Island Cocky Apple

Registration #	Variety	Native Location	Variety	Specimen	Genome
727.153.104.070.1	*sundaica*	E. Indies (Borneo, Java, Sulawesi, Sumatra)	Original	Greater Sunda Island	Cocky Apple
727.153.104.070.2	*sumatrana*	E. Indies (Sumatra)	Sumatra	Greater Sunda Island	Cocky Apple

727.153.104.08 **SPECIES 08** *timorensis* Lesser Sunda Island Cocky Apple

Registration #	Variety	Native Location	Variety	Specimen	Genome
727.153.104.080.1	*timorensis*	E. Indies (Java to Timor)	Original	Lesser Sunda Island	Cocky Apple
727.153.104.080.2	*tetraptera*	E. Indies (Java to Timor)	Four Wing	Lesser Sunda Island	Cocky Apple

727.153.104.09 **SPECIES 09** *valida* East Malaysia Cocky Apple

Registration #	Variety	Native Location	Variety	Specimen	Genome
727.153.104.090.1	*valida*	E. Malaysia, E. Indies (Malay Is.)	Original	East Malaysia	Cocky Apple
727.153.104.090.2	*alata*	E. Malaysia, E. Indies (Malay Is.)	Winged	East Malaysia	Cocky Apple
727.153.104.090.3	*elliptica*	E. Malaysia, E. Indies (Malay Is.)	Elliptic Leaf	East Malaysia	Cocky Apple
727.153.104.090.4	*forbesii*	E. Malaysia, E. Indies (Malay Is.)	Forbes	East Malaysia	Cocky Apple
727.153.104.090.5	*littoralis*	E. Malaysia, E. Indies (Malay Is.)	Seaside	East Malaysia	Cocky Apple
727.153.104.090.6	*undulata*	E. Malaysia, E. Indies (Malay Is.)	Wavy	East Malaysia	Cocky Apple

Violet Class (*Violopsida*)

727.16 Class 16 *Violopsida* Violet

Violet Order (*Violales*)

Order 1 *Violales* Violet

Papaya Family (*Caricaceae*)

Family 4 *Caricaceae* Papaya

The Complete Book of Fruits, ISBN:0670-86752-7, page, 153, states that "The fruits uncannily resemble melons, turning yellow-orange as the ripen. The flesh is usually pink with a central hole full of small round seeds and can weigh up to 5 lb. Although indigenous to Central America, the papaya has rapidly spread to every warm country."

PAPAYA (*Carica*)

GENUS 01 *Carica* Papaya
727.161.401.01 **SPECIES 01** *africana* African Papaya

Registration #	Variety	Native Location	Variety	Specimen	Genome
727.161.401.010.1	*africana*	Africa	Original	African	Papaya
727.161.401.010.2	*zairensis*	Africa	Zaire Dwarf	African	Papaya

727.161.401.02 **SPECIES 02** *cariflora* Flowering Papaya
727.161.401.03 **SPECIES 03** *formosa* Giant Red Papaya
727.161.401.04 **SPECIES 04** *goudetiana* Goudet Papaya
727.161.401.05 **SPECIES 05** *indica* India Papaya

Registration #	Variety	Native Location	Variety	Specimen	Genome
727.161.401.050.1	*indica*	S. Asia (India)	Original	India	Papaya
727.161.401.050.2	*honeydewii*	S. Asia (India)	Honeydew	India	Papaya
727.161.401.050.3	*pusa-nahai*	S. Asia (India)	Pusa Naha Dwarf	India	Papaya

727.161.401.06 **SPECIES 06** *mexicana* Mexican Papaya
727.161.401.07 **SPECIES 07** *monoica* Andes Mountain Papaya

Registration #	Variety	Native Location	Variety	Specimen	Genome
727.161.401.070.1	*monoica*	S. America (E. Andes Mtns.)	Original	Andes Mountain	Papaya
727.161.401.070.2	*peladerai*	S. America (E. Andes Mtns.)	Peladera	Andes Mountain	Papaya
727.161.401.070.3	*red-mountainii*	S. America (E. Andes Mtns.)	Red	Andes Mountain	Papaya
727.161.401.070.4	*toronchei*	S. America (E. Andes Mtns.)	Toronche	Andes Mountain	Papaya

727.161.401.08 **SPECIES 08** *papaya* Garden Papaya

Registration #	Variety	Native Location	Variety	Specimen	Genome
727.161.401.080.1	*papaya*	S. America (Tropical Lowlands)	Original	Garden	Papaya

Registration #	Variety	Native Location	Variety	Specimen	Genome
727.161.401.080.2	*nana*	S. America (Tropical Lowlands)	Dwarf	Garden	Papaya

727.161.401.09 **SPECIES 08** *parviflora* Central America Mountain Papaya
727.161.401.10 **SPECIES 10** *pentagona* South America Mountain Papaya

Registration #	Variety	Native Location	Variety	Specimen	Genome
727.161.401.100.1	*pentagona*	S. America (Colombia & Ecuador Highlands)	Original	South America Mountain	Papaya
727.161.401.100.2	*babacoi*	S. America (Colombia & Ecuador Highlands)	Babaco	South America Mountain	Papaya
727.161.401.100.3	*x heilbornii*	S. America (Colombia & Ecuador Highlands)	Heilborn	South America Mountain	Papaya

727.161.401.11 **SPECIES 11** *pubescens* Downy Mountain Papaya

Registration #	Variety	Native Location	Variety	Specimen	Genome
727.161.401.110.1	*pubescens*	S. America (Tropical Highlands)	Original	Downy Mountain	Papaya
727.161.401.110.2	*candamarcensis*	S. America (Tropical Highlands)	Candamarc	Downy Mountain	Papaya
727.161.401.110.3	*chamburrosensis*	S. America (Tropical Highlands)	Chamburo	Downy Mountain	Papaya

727.161.401.12 **SPECIES 12** *quercifolia* Oak Leaf Mountain Papaya
727.161.401.13 **SPECIES 13** *soloi* Solo Papaya

Registration #	Variety	Native Location	Variety	Specimen	Genome
727.161.401.130.1	*soloi*	America (Tropical)	Original	Solo	Papaya
727.161.401.130.2	*hawaiiensis*	Pac. Is. (Hawaiian Is.)	Hawaiian	Solo	Papaya
727.161.401.130.3	*sunrisei*	America (Tropical)	Sunrise	Solo	Papaya
727.161.401.130.4	*sunsetii*	America (Tropical)	Sunset	Solo	Papaya
727.161.401.130.5	*waimanolo-nana*	Pac. Is. (Hawaiian Is.)	Waimanolo Dwarf	Solo	Papaya

727.161.401.14 **SPECIES 14** *stipulata* Siglalón Papaya
727.161.401.15 **SPECIES 15** *taiwanensis* Taiwan Papaya

Registration #	Variety	Native Location	Variety	Specimen	Genome
727.161.401.150.1	*taiwanensis*	E. Asia (Taiwan)	Original	Taiwan	Papaya
727.161.401.150.2	*known-youi-1*	E. Asia (Taiwan)	Known You #1	Taiwan	Papaya
727.161.401.150.3	*red-ladyi*	E. Asia (Taiwan)	Red Lady Dwarf	Taiwan	Papaya
727.161.401.150.4	*red-ladyi-xps-15*	E. Asia (Taiwan)	Red Lady Dwarf XPS #15	Taiwan	Papaya

727.161.401.16 **SPECIES 16** *thailandica* Thai Papaya

Registration #	Variety	Native Location	Variety	Specimen	Genome
727.161.401.160.1	*thailandica*	S. Asia (Thailand)	Original	Thai	Papaya
727.161.401.160.2	*gigantea*	S. Asia (Thailand)	Giant	Thai	Papaya
727.161.401.160.3	*nana*	S. Asia (Thailand)	Dwarf	Thai	Papaya
727.161.401.160.4	*rubra-aurantia*	S. Asia (Thailand)	Red Orange	Thai	Papaya

JARILL PAPAYA (*Jarilla*)

GENUS 02 *Jarilla* Jarill Papaya
727.161.402.01 **SPECIES 01** *heterophylla* Jarill Papaya

Registration #	Variety	Native Location	Variety	Specimen	Genome
727.161.402.010.1	*heterophylla*	.	Original	Variable Leaf	Jarill Papaya
727.161.402.010.2	*chocola*	.	Chocola	Variable Leaf	Jarill Papaya

GENUS *Papaya* (Synonym for *Carica*)

Melon Family (*Cucurbitaceae*)

Family 6 *Cucurbitaceae* Melon

GOURD MELON (*Benincasa*)

Stearn's Dictionary of Plant Names for Gardeners, ISBN:0-88192-556-X, page 63, states that "Annual edible squash like herbs with edible fruits, named for Count Giuseppe Benincasa (died 1596) Italian botanist who founded the Botanic Garden in Pisa [Italy]".

GENUS 01 *Benincasa* Gourd Melon
727.161.601.01 **SPECIES 01** *hispida* Gourd Melon

Registration #	Variety	Native Location	Variety	Specimen	Genome
727.161.601.010.1	*hispida*	Pac. Is.	Original	Hairy	Gourd Melon
727.161.601.010.2	*cerifera*	Pac. Is.	Wax Gourd	Hairy	Gourd Melon

WATERMELON (*Citrullus*)

Fruits and Vegetables of the World, ISBN:0-681-21883-5, page 24, states that "Some watermelons weighing up to 40 pounds grow only in very hot regions". Fruits and Vegetables of the World, ISBN:0-681-21883-5, page 25, states that "In general Mediterranean watermelons are round, while American watermelons are oval-shaped. There is a variety cultivated in Thailand that has a bright yellow skin". Flora, ISBN:1-74048-097-X, page 383 states that for *Citrullus lanatus* "Annual climber or trailer from Namibia, naturalized elsewhere". In the listings below all American watermelons are oblong oval unless otherwise noted. In the listings below, all of the European, Asian, and African watermelons are round unless otherwised noted.

GENUS 02 *Citrullus* Watermelon

727.161.602.01 **SPECIES 01** *colocynthis* Colocynth Watermelon

727.161.602.02 **SPECIES 02** *lanatus-alba-parva* Small (20lb) White Watermelon

Registration #	Variety	Native Location	Variety	Specimen	Genome
727.161.602.020.1	*alba-parva*	USA	Original	Small (20lb) White	Watermelon
727.161.602.020.2	*curtis-showell-white-fleshii*	USA (MO)	Curtis Showell	Small (20lb) White	Watermelon

727.161.602.03 **SPECIES 03** *lanatus-alba-mini* Mini (10lb) White Watermelon

Registration #	Variety	Native Location	Variety	Specimen	Genome
727.161.602.030.1	*alba-mini*	Africa, Asia	Original	Mini (10lb) White	Watermelon
727.161.602.030.2	*cream-of-saskatchewanii*	Asia (Saskatchewan, Russia)	Cream Of Saskatchewan Round	Mini (10lb) White	Watermelon
727.161.602.030.3	*kenyaensis*	Africa (Kenya)	Kenya	Mini (10lb) White	Watermelon

727.161.602.04 **SPECIES 04** *lanatus-alba-micro* Tiny (5-8lb) White Watermelon

Registration #	Variety	Native Location	Variety	Specimen	Genome
727.161.602.040.1	*alba-micro*	Asia	Original	Tiny (5-8lb) White	Watermelon
727.161.602.040.2	*japanese-cream-suikai*	E. Asia (Japan)	Japanese Suika	Tiny (5-8lb) White	Watermelon
727.161.602.040.3	*white-wonderi*	E. Asia (Japan)	White Wonder	Tiny (5-8lb) White	Watermelon

727.161.602.05 **SPECIES 05** *lanatus-alba-nano* Wee (2-4lb) White Watermelon

Registration #	Variety	Native Location	Variety	Specimen	Genome
727.161.602.050.1	*alba-nano*	Asia	Original	Wee (2-4lb) White	Watermelon
727.161.602.050.2	*white-meatedii*	E. Asia (Japan)	White Meated	Wee (2-4lb) White	Watermelon

727.161.602.06 **SPECIES 06** *lanatus-aurantia-larga* Large (40lb) Orange Watermelon

Registration #	Variety	Native Location	Variety	Specimen	Genome
727.161.602.060.1	*aurantiaca-larga*	USA	Original	Large (40lb) Orange	Watermelon
727.161.602.060.2	*orangegloi*	USA (MO)	Orange Glow	Large (40lb) Orange	Watermelon
727.161.602.060.3	*tender-sweet-orange-fleshii*	USA (MO)	Tender Sweet	Large (40lb) Orange	Watermelon

727.161.602.07 **SPECIES 07** *lanatus-aurantia-media* Medium (30lb) Orange Watermelon

727.161.602.08 **SPECIES 08** *lanatus-aurantia-nominalis* Nominal (25lb) Orange Watermelon

Registration #	Variety	Native Location	Variety	Specimen	Genome
727.161.602.080.1	*aurantia-nominalis*	USA	Original	Nominal (25lb) Orange	Watermelon
727.161.602.080.2	*lucious-goldenii*	USA (IA)	Luscious Golden	Nominal (25lb) Orange	Watermelon

727.161.602.09 **SPECIES 09** *lanatus-aurantia-parva* Small (20lb) Orange Watermelon

Registration #	Variety	Native Location	Variety	Specimen	Genome
727.161.602.090.1	*aurantia-parva*	S. America	Original	Small (20lb) Orange	Watermelon
727.161.602.090.2	*chilean-black-seededii*	S. Amer. (Chile)	Chilean Black Seeded	Small (20lb) Orange	Watermelon

727.161.602.10 **SPECIES 10** *lanatus-aurantia-mini* Mini (10lb) Orange Watermelon

727.161.602.11 **SPECIES 11** *lanatus-aurantia-micro* Tiny (5-8lb) Orange Watermelon

Registration #	Variety	Native Location	Variety	Specimen	Genome
727.161.602.110.1	*aurantia-micro*	Asia	Original	Tiny (5-8lb) Orange	Watermelon
727.161.602.110.2	*sweet-siberiana*	Asia (Siberia)	Sweet Siberian	Tiny (5-8lb) Orange	Watermelon

727.161.602.12 **SPECIES 12** *lanatus-aurantia-helva-media* Medium (30lb) Orange Yellow Watermelon

Registration #	Variety	Native Location	Variety	Specimen	Genome
727.161.602.120.1	*aurantia-helva-media*	USA	Original	Medium (30lb) Orange Yellow	Watermelon
727.161.602.120.2	*golden-jubileei*	USA (FL)	Golden Jubilee	Medium (30lb) Orange Yellow	Watermelon
727.161.602.120.3	*tendersweetii*	USA (MO)	Tender Sweet	Medium (30lb) Orange Yellow	Watermelon

727.161.602.13 **SPECIES 13** *lanatus-aurantia-rubra-larga* Large (40lb) Orange Red Watermelon

Registration #	Variety	Native Location	Variety	Specimen	Genome
727.161.602.130.1	*aurantia-rubra-larga*	USA	Original	Large (40lb) Orange Red	Watermelon
727.161.602.130.2	*dixie-queenii*	USA (FL)	Dixie Queen Blunt	Large (40lb) Orange Red	Watermelon
727.161.602.130.3	*dixie-queenii-wilt-resistantii*	USA (FL)	Dixie Queen Wilt Resistant	Large (40lb) Orange Red	Watermelon

727.161.602.14 **SPECIES 14** *lanatus-aurantia-rubra-media* Medium (30lb) Orange Red Watermelon

Registration #	Variety	Native Location	Variety	Specimen	Genome
727.161.602.140.1	*aurantia-rubra-media*	USA	Original	Medium (30lb) Orange Red	Watermelon
727.161.602.140.2	*colorado-striped-tarahumarai*	USA (CO)	Colorado Striped Tarahumara	Medium (30lb) Orange Red	Watermelon

727.161.602.15 **SPECIES 15** *lanatus-aurantia-rubra-parva* Small (20lb) Orange Red Watermelon

Registration #	Variety	Native Location	Variety	Specimen	Genome
727.161.602.150.1	*aurantia-rubra-parva*	USA	Original	Small (20lb) Orange Red	Watermelon
727.161.602.150.2	*chris-crossi*	USA (IA)	Chris Cross	Small (20lb) Orange Red	Watermelon
727.161.602.150.3	*peacockii-improva*	USA (CA)	Peacock Improved	Small (20lb) Orange Red	Watermelon
727.161.602.150.4	*peacockii-picnici*	USA (IA)	Peacock Picnic	Small (20lb) Orange Red	Watermelon

727.161.602.16 **SPECIES 16** *lanatus-aurantia-rubra-mini* Mini (10lb) Orange Red Watermelon

Registration #	Variety	Native Location	Variety	Specimen	Genome
727.161.602.160.1	*aurantia-rubra-mini*	USA	Original	Mini (10lb) Orange Red	Watermelon
727.161.602.160.2	*blacktail-mountainii*	USA (OR)	Blacktail Mountain Round	Mini (10lb) Orange Red	Watermelon

Registration #	Variety	Native Location	Variety	Specimen	Genome
727.161.602.160.3	*sugar-babyi-bushii*	USA (OR)	Sugar Baby Bush	Mini (10lb) Orange Red	Watermelon
727.161.602.160.4	*sweet-northern-prizei*	USA (NM)	Sweet Northern Prize	Mini (10lb) Orange Red	Watermelon
727.161.602.160.5	*swift-watersii*	Canada (ON)	Swift Waters	Mini (10lb) Orange Red	Watermelon
727.161.602.17	**SPECIES 17** *lanatus-aurantia-rubra-micro*		Tiny (5-8lb) Orange Red Watermelon		
Registration #	Variety	Native Location	Variety	Specimen	Genome
727.161.602.170.1	*aurantia-rubra-micro*	USA	Original	Tiny (5-8lb) Orange Red	Watermelon
727.161.602.170.2	*sugar-babyi*	USA	Sugar Baby	Tiny (5-8lb) Orange Red	Watermelon
727.161.602.170.3	*sugar-bushii-burpeei*	USA (PA)	Sugar Bush Burpee	Tiny (5-8lb) Orange Red	Watermelon
727.161.602.18	**SPECIES 18** *lanatus-citroides*		Citron Watermelon		
Registration #	Variety	Native Location	Variety	Specimen	Genome
727.161.602.180.1	*citroides*	USA	Original	Citron	Watermelon
727.161.602.180.2	*californica*	USA	California 3lb	Citron	Watermelon
727.161.602.180.3	*green-seededii*	USA	Green Seeded 11lb	Citron	Watermelon
727.161.602.180.4	*red-seededii*	USA	Red Seeded 11lb	Citron	Watermelon
727.161.602.180.5	*red-shinei*	USA (OR)	Red Shine 8lb	Citron	Watermelon
727.161.602.19	**SPECIES 19** *lanatus-helva-gigantea*		Giant (60lb) Yellow Watermelon		
Registration #	Variety	Native Location	Variety	Specimen	Genome
727.161.602.190.1	*helva-gigantea*	USA	Original	Giant (60lb) Yellow	Watermelon
727.161.602.190.2	*black-diamond-yellow-fleshii*	USA (FL)	Black Diamond	Giant (60lb) Yellow	Watermelon
727.161.602.20	**SPECIES 20** *lanatus-helva-larga*		Large (40lb) Yellow Watermelon		
727.161.602.21	**SPECIES 21** *lanatus-helva-media*		Medium (30lb) Yellow Watermelon		
Registration #	Variety	Native Location	Variety	Specimen	Genome
727.161.602.210.1	*helva-media*	USA, Mexico	Original	Medium (30lb) Yellow	Watermelon
727.161.602.210.2	*tendersweet-yellow-fleshii*	USA (SC)	Tender Sweet	Medium (30lb) Yellow	Watermelon
727.161.602.210.3	*tohono-o'odham-yellow-meatedii*	Mexico	Tohono O'odham	Medium (30lb) Yellow	Watermelon
727.161.602.22	**SPECIES 22** *lanatus-helva-nominalis*		Nominal (25lb) Yellow Watermelon		
Registration #	Variety	Native Location	Variety	Specimen	Genome
727.161.602.220.1	*helva-nominalis*	USA	Original	Nominal (25lb) Yellow	Watermelon
727.161.602.220.2	*desert-kingii-yellow-fleshii*	USA	Desert King	Nominal (25lb) Yellow	Watermelon
727.161.602.220.3	*golden-honeyi*	USA (NE)	Golden Honey Round	Nominal (25lb) Yellow	Watermelon
727.161.602.220.4	*mountain-sweet-yellowii*	USA (NE)	Mountain Sweet Yellow	Nominal (25lb) Yellow	Watermelon
727.161.602.220.5	*sugar-lump-yellowii*	USA	Sugar Lump	Nominal (25lb) Yellow	Watermelon
727.161.602.220.6	*tendergoldii*	USA (NM)	Tender Gold	Nominal (25lb) Yellow	Watermelon
727.161.602.220.7	*tendergoldii-willhitei*	USA (TX)	Tender Gold Willhite	Nominal (25lb) Yellow	Watermelon
727.161.602.220.8	*yellow-crimsonii*	USA (IN)	Yellow Crimson	Nominal (25lb) Yellow	Watermelon
727.161.602.23	**SPECIES 23** *lanatus-helva-parva*		Small (20lb) Yellow Watermelon		
Registration #	Variety	Native Location	Variety	Specimen	Genome
727.161.602.230.1	*helva-parva*	USA	Original	Small (20lb) Yellow	Watermelon
727.161.602.230.2	*au-golden-produceri*	USA (AL)	Au-Golden Producer Blocky	Small (20lb) Yellow	Watermelon
727.161.602.230.3	*moon-&-starsii-yellow-fleshii*	USA (GA)	Moon & Stars	Small (20lb) Yellow	Watermelon
727.161.602.230.4	*tastigoldii*	USA (TX)	Tastey Gold Round	Small (20lb) Yellow	Watermelon
727.161.602.24	**SPECIES 24** *lanatus-helva-mini*		Mini (10lb) Yellow Watermelon		
Registration #	Variety	Native Location	Variety	Specimen	Genome
727.161.602.240.1	*helva-mini*	USA	Original	Mini (10lb) Yellow	Watermelon
727.161.602.240.2	*yellow-shipperi*	USA (CO)	Yellow Shipper	Mini (10lb) Yellow	Watermelon
727.161.602.25	**SPECIES 25** *lanatus-helva-micro*		Tiny (5-8lb) Yellow Watermelon		
Registration #	Variety	Native Location	Variety	Specimen	Genome
727.161.602.250.1	*helva-micro*	USA	Original	Tiny (5-8lb) Yellow	Watermelon
727.161.602.250.2	*early-moonbeamii*	N. USA	Early Moonbeam	Tiny (5-8lb) Yellow	Watermelon
727.161.602.250.3	*gold-babyi*	USA (CA)	Gold Baby	Tiny (5-8lb) Yellow	Watermelon
727.161.602.250.4	*hopii-yellow-meatedii*	USA (AZ)	Hopi	Tiny (5-8lb) Yellow	Watermelon
727.161.602.250.5	*petite-yellowii*	USA	Petite Yellow	Tiny (5-8lb) Yellow	Watermelon
727.161.602.26	**SPECIES 26** *lanatus-pinca-larga*		Large (40lb) Pink Watermelon		
Registration #	Variety	Native Location	Variety	Specimen	Genome
727.161.602.260.1	*pinca-larga*	USA	Original	Large (40lb) Pink	Watermelon
727.161.602.260.2	*garrison-rattlesnakei*	USA (GA)	Garrison Rattlesnake	Large (40lb) Pink	Watermelon
727.161.602.27	**SPECIES 27** *lanatus-pinca-media*		Medium (30lb) Pink Watermelon		
Registration #	Variety	Native Location	Variety	Specimen	Genome
727.161.602.270.1	*pinca-media*	USA	Original	Medium (30lb) Pink	Watermelon
727.161.602.270.2	*arikara*	USA (WA)	Arikara	Medium (30lb) Pink	Watermelon
727.161.602.270.3	*georgica-giantii-stripedii*	USA (GA)	Georgia Striped	Medium (30lb) Pink	Watermelon
727.161.602.270.4	*moon-&-starsi-north-stari*	USA (IL)	Moon & Stars North Star	Medium (30lb) Pink	Watermelon
727.161.602.270.5	*wonder-melonii*	USA	Wonder Melon	Medium (30lb) Pink	Watermelon

727.161.602.28 **SPECIES 28** *lanatus-pinca-nominalis* Nominal (25lb) Pink Watermelon

Registration #	Variety	Native Location	Variety	Specimen	Genome
727.161.602.280.1	*pinca-nominalis*	USA	Original	Nominal (25lb) Pink	Watermelon
727.161.602.280.2	*early-arizonai*	USA (AZ)	Early Arizona	Nominal (25lb) Pink	Watermelon
727.161.602.280.3	*moon-&-starsii-pink-flesh-amishii*	USA (PA)	Moon & Stars Amish	Nominal (25lb) Pink	Watermelon
727.161.602.280.4	*nancyi*	USA (GA)	Nancy	Nominal (25lb) Pink	Watermelon
727.161.602.280.5	*navajo-red-seededii*	USA (AZ)	Navaho Red Seeded	Nominal (25lb) Pink	Watermelon
727.161.602.280.6	*sweet-princessi*	USA (NC)	Sweet Princess	Nominal (25lb) Pink	Watermelon

727.161.602.29 **SPECIES 29** *lanatus-pinca-parva* Nominal (25lb) Pink Watermelon

Registration #	Variety	Native Location	Variety	Specimen	Genome
727.161.602.290.1	*pinca-parva*	USA	Original	Small (20lb) Pink	Watermelon
727.161.602.290.2	*astrakhanskii*	Asia (Russia)	Astrakhansky	Small (20lb) Pink	Watermelon
727.161.602.290.3	*chelsea*	USA	Chelsea	Small (20lb) Pink	Watermelon
727.161.602.290.4	*georgica-rattlesnakei*	USA (GA)	Georgia Rattlesnake	Small (20lb) Pink	Watermelon
727.161.602.290.5	*king-&-queenii*	USA	King & Queen Round	Small (20lb) Pink	Watermelon
727.161.602.290.6	*strawberryi*	USA (VA)	Strawberry	Small (20lb) Pink	Watermelon

727.161.602.30 **SPECIES 30** *lanatus-pinca-mini* Mini (10lb) Pink Watermelon

Registration #	Variety	Native Location	Variety	Specimen	Genome
727.161.602.300.1	*pinca-mini*	USA, Africa, M. East	Original	Mini (10lb) Pink	Watermelon
727.161.602.300.2	*early-yatesii*	USA (IA)	Early Yates	Mini (10lb) Pink	Watermelon
727.161.602.300.3	*graybellei*	M. East (Israel)	Graybelle	Mini (10lb) Pink	Watermelon
727.161.602.300.4	*osh-kirgizia*	USA (IA)	Osh Kirgiz	Mini (10lb) Pink	Watermelon
727.161.602.300.5	*quetzalii*	Mexico	Quetzal	Mini (10lb) Pink	Watermelon
727.161.602.300.6	*will's-sugari*	USA (IA)	Will's Sugar	Mini (10lb) Pink	Watermelon
727.161.602.300.7	*texas-pinkii*	USA (TX)	Texas Blocky	Mini (10lb) Pink	Watermelon
727.161.602.300.8	*triplesweet-seedlessi*	Africa (Namibia)	Triplesweet Seedless	Mini (10lb) Pink	Watermelon

727.161.602.31 **SPECIES 31** *lanatus-pinca-micro* Tiny (5-8lb) Pink Watermelon

Registration #	Variety	Native Location	Variety	Specimen	Genome
727.161.602.310.1	*pinca-micro*	USA, Africa, M. East	Original	Tiny (5-8lb) Pink	Watermelon
727.161.602.310.2	*fordhook-hybridii*	Africa (Namibia)	Fordhook Hybrid	Tiny (5-8lb) Pink	Watermelon
727.161.602.310.3	*Malalii*	M. East (Israel)	Malali Oblong	Tiny (5-8lb) Pink	Watermelon
727.161.602.310.4	*petite-sweetii*	USA (KS)	Petite Sweet Blocky	Tiny (5-8lb) Pink	Watermelon

727.161.602.32 **SPECIES 32** *lanatus-rubra-enorma* Enormous (120lb) Red Watermelon

Registration #	Variety	Native Location	Variety	Specimen	Genome
727.161.602.320.1	*rubra-enorma*	USA	Original	Enormous (120lb) Red	Watermelon
727.161.602.320.2	*carolina-crossii*	USA (SC)	Carolina Cross	Enormous (120lb) Red	Watermelon
727.161.602.320.3	*carolina-crossii-183*	USA (SC)	Carolina Cross #183	Enormous (120lb) Red	Watermelon
727.161.602.320.4	*carolina-crossii-200*	USA (SC)	Carolina Cross #200	Enormous (120lb) Red	Watermelon
727.161.602.320.5	*cobb-gemii*	USA	Cobb Gem	Enormous (120lb) Red	Watermelon
727.161.602.320.6	*kolb's-gemii*	USA	Kolb's Gem	Enormous (120lb) Red	Watermelon
727.161.602.320.7	*weeks-n.-carolina-giantii*	USA (NC)	Weeks N. Carolina Giant	Enormous (120lb) Red	Watermelon

727.161.602.33 **SPECIES 33** *lanatus-rubra-gigantea* Giant (60lb) Red Watermelon

Registration #	Variety	Native Location	Variety	Specimen	Genome
727.161.602.330.1	*rubra-gigantea*	USA	Original	Giant (60lb) Red	Watermelon
727.161.602.330.2	*mountain-hoosieri*	USA (IN)	Mountain Hoosier	Giant (60lb) Red	Watermelon
727.161.602.330.3	*tom-watsonii-white-seededii*	USA (PA)	Tom Watson White Seeded	Giant (60lb) Red	Watermelon

727.161.602.34 **SPECIES 34** *lanatus-rubra-larga* Large (40lb) Red Watermelon

Registration #	Variety	Native Location	Variety	Specimen	Genome
727.161.602.340.1	*rubra-larga*	USA	Original	Large (40lb) Red	Watermelon
727.161.602.340.2	*au-jubilantii*	USA (AL)	Au Jubilant	Large (40lb) Red	Watermelon
727.161.602.340.3	*big-crimsonii*	USA (TX)	Big Crimson	Large (40lb) Red	Watermelon
727.161.602.340.4	*black-diamondii*	USA (FL)	Black Diamond	Large (40lb) Red	Watermelon
727.161.602.340.5	*black-diamondii-yellow-bellyi*	USA (FL)	Black Diamond Yellow Belly	Large (40lb) Red	Watermelon
727.161.602.340.6	*candy-redii*	Australia	Candy Red	Large (40lb) Red	Watermelon
727.161.602.340.7	*congoensis*	Africa (Congo)	Congo Long	Large (40lb) Red	Watermelon
727.161.602.340.8	*florida-giant-cannonballii*	USA (FL)	Florida Giant Cannonball	Large (40lb) Red	Watermelon
727.161.602.340.9	*moon-&-starsi-van-dorenii*	USA (MO)	Moon & Stars Van Doren	Large (40lb) Red	Watermelon
727.161.602.341.0	*texas-giantii*	USA (TX)	Texas Giant	Large (40lb) Red	Watermelon
727.161.602.341.1	*warpaintii*	USA (WA)	War Paint	Large (40lb) Red	Watermelon

727.161.602.35 **SPECIES 35** *lanatus-rubra-media* Medium (30lb) Red Watermelon

Registration #	Variety	Native Location	Variety	Specimen	Genome
727.161.602.350.1	*rubra-media*	W. Indies	Original	Medium (30lb) Red	Watermelon
727.161.602.350.2	*calsweetii*	USA (CA)	Calsweet Blocky	Medium (30lb) Red	Watermelon
727.161.602.350.3	*calsweetii-classici*	USA (CA)	Calsweet Classic	Medium (30lb) Red	Watermelon

Registration #	Variety	Native Location	Variety	Specimen	Genome
727.161.602.350.4	*charleston-elitei*	USA (SC)	Charleston Elite	Medium (30lb) Red	Watermelon
727.161.602.350.5	*charleston-grayi*	USA (SC)	Charleston Gray	Medium (30lb) Red	Watermelon
727.161.602.350.6	*charleston-grayi-5*	USA (SC)	Charleston Gray #5	Medium (30lb) Red	Watermelon
727.161.602.350.7	*charleston-grayi-133*	USA (SC)	Charleston Gray #133	Medium (30lb) Red	Watermelon
727.161.602.350.8	*chubby-grayi*	USA (TX)	Chubby Gray Blocky	Medium (30lb) Red	Watermelon
727.161.602.350.9	*fairfaxii*	USA (SC)	Fairfax	Medium (30lb) Red	Watermelon
727.161.602.351.0	*iopridei*	USA (IA)	Iowa Pride	Medium (30lb) Red	Watermelon
727.161.602.351.1	*irish-grayi*	USA (OH)	Irish Gray	Medium (30lb) Red	Watermelon
727.161.602.351.2	*jubileei*	USA (SC)	Jubilee	Medium (30lb) Red	Watermelon
727.161.602.351.3	*jubileei-2*	USA (SC)	Jubilee 2	Medium (30lb) Red	Watermelon
727.161.602.351.4	*jubileei-improva*	USA (FL)	Jubilee Improved	Medium (30lb) Red	Watermelon
727.161.602.351.5	*jubileei-registeredii*	USA (PA)	Jubilee Registered	Medium (30lb) Red	Watermelon
727.161.602.351.6	*juliettii-NV4317*	USA (NC)	Juliett	Medium (30lb) Red	Watermelon
727.161.602.351.7	*kleckleyi-resistantii-6*	USA (IA)	Kleckley Risistant #6	Medium (30lb) Red	Watermelon
727.161.602.351.8	*kleckleyi-sweetii*	USA (IA)	Kleckley Sweet	Medium (30lb) Red	Watermelon
727.161.602.351.9	*kleckleyi-sweetii-improvedii*	USA (IA)	Kleckley Sweet Improved	Medium (30lb) Red	Watermelon
727.161.602.352.0	*moon-&-starsi-long-milky-wayi*	USA (CA)	Moon & Stars Milky Way	Medium (30lb) Red	Watermelon
727.161.602.352.1	*moon-&-starsi-cherokeei*	USA (IA)	Cherokee Moon & Stars	Medium (30lb) Red	Watermelon
727.161.602.352.2	*snakeskin-bushii*	USA (NM)	Snakeskin Bush	Medium (30lb) Red	Watermelon
727.161.602.352.3	*stone-mountainii*	USA (GA)	Stone Mountain	Medium (30lb) Red	Watermelon
727.161.602.352.4	*stone-mountainii-improva-5*	USA (GA)	Stone Mountain Improved #5	Medium (30lb) Red	Watermelon
727.161.602.352.5	*tom-watsonii*	USA (PA)	Tom Watson	Medium (30lb) Red	Watermelon
727.161.602.352.6	*veronai*	USA (NM)	Verona Black Diamond	Medium (30lb) Red	Watermelon
727.161.602.36	**SPECIES 36** *lanatus-rubra-nominalis*		Nominal (25lb) Red Watermelon		
Registration #	**Variety**	**Native Location**	**Variety**	**Specimen**	**Genome**
727.161.602.360.1	*rubra-nominalis*	USA	Original	Nominal (25lb) Red	Watermelon
727.161.602.360.2	*acomai*	USA (AZ)	Acoma	Nominal (25lb) Red	Watermelon
727.161.602.360.3	*ali-babai*	M. East	Ali Baba	Nominal (25lb) Red	Watermelon
727.161.602.360.4	*art-combe-red-seededii*	USA (AZ)	Art Combe Red Seeded	Nominal (25lb) Red	Watermelon
727.161.602.360.5	*crimson-sweetii*	USA (KS)	Crimson Sweet Blocky	Nominal (25lb) Red	Watermelon
727.161.602.360.6	*dixie-leei*	USA (FL)	Dixie Lee Round	Nominal (25lb) Red	Watermelon
727.161.602.360.7	*fistulosus*	S. Asia (India)	Gourd Tinda	Nominal (25lb) Red	Watermelon
727.161.602.360.8	*giza*	M. East	Giza	Nominal (25lb) Red	Watermelon
727.161.602.360.9	*jemezii*	S. USA	Jemez	Nominal (25lb) Red	Watermelon
727.161.602.361.0	*jubileei-FFSS*	USA (OH)	Jubilee FFSS	Nominal (25lb) Red	Watermelon
727.161.602.361.1	*jumanosa*	USA (TX)	Jumanos	Nominal (25lb) Red	Watermelon
727.161.602.361.2	*karpas-greekii*	Europe (Greece)	Karpas Greek	Nominal (25lb) Red	Watermelon
727.161.602.361.3	*klondykei-RS57*	USA (CA)	Klondyke Resistant Shipper #57	Nominal (25lb) Red	Watermelon
727.161.602.361.4	*klondykei-striped-blue-ribbonii*	USA (CA)	Klondyke Striped Blue Ribbon	Nominal (25lb) Red	Watermelon
727.161.602.361.5	*long-crimsonii*	USA (TX)	Long Crimson Blocky	Nominal (25lb) Red	Watermelon
727.161.602.361.6	*louisiana-sweetii*	USA (LA)	Louisiana Sweet	Nominal (25lb) Red	Watermelon
727.161.602.361.7	*mayoi*	Mexico (Sinaloa)	Mayo	Nominal (25lb) Red	Watermelon
727.161.602.361.8	*mayoi-sandia*	Mexico (Sinaloa)	Mayo Sandia	Nominal (25lb) Red	Watermelon
727.161.602.361.9	*new-queenii*	Africa (Namibia)	New Queen	Nominal (25lb) Red	Watermelon
727.161.602.362.0	*odemii*	M. East (Israel)	Odem Oval	Nominal (25lb) Red	Watermelon
727.161.602.362.1	*peacockii-stripedii*	USA (CA)	Peacock Striped	Nominal (25lb) Red	Watermelon
727.161.602.362.2	*rainbowii*	USA (ID)	Rainbow	Nominal (25lb) Red	Watermelon
727.161.602.362.3	*rio-mayo-sakobari*	Mexico (Chihuahua)	Rio Mayo Sakobar	Nominal (25lb) Red	Watermelon
727.161.602.362.4	*rio-mayo-san-miguelii*	Mexico (Chihuahua)	Rio Mayo Miguel	Nominal (25lb) Red	Watermelon
727.161.602.362.5	*san-juanii*	USA (AZ)	San Juan	Nominal (25lb) Red	Watermelon
727.161.602.362.6	*santo-domingo-winteri*	Mexico (Santo Domingo)	Santo Domingo Winter Round	Nominal (25lb) Red	Watermelon
727.161.602.362.7	*stars-&-stripesii*	USA (ND)	Stars And Stripes	Nominal (25lb) Red	Watermelon
727.161.602.362.8	*thai-blackii*	S. Asia (Thailand)	Thai Black	Nominal (25lb) Red	Watermelon
727.161.602.362.9	*unclei-E*	USA (MS)	Uncle E Round	Nominal (25lb) Red	Watermelon
727.161.602.37	**SPECIES 37** *lanatus-rubra-parva*		Small (20lb) Red Watermelon		
Registration #	**Variety**	**Native Location**	**Variety**	**Specimen**	**Genome**
727.161.602.370.1	*rubra-parva*	USA	Original	Small (20lb) Red	Watermelon
727.161.602.370.2	*au-produceri*	USA (AL)	Au Producer	Small (20lb) Red	Watermelon
727.161.602.370.3	*au-sweet-scarletii*	USA (AL)	Au Sweet Scarlet Globe	Small (20lb) Red	Watermelon
727.161.602.370.4	*blackstonei*	USA (Florida)	Blackstone	Small (20lb) Red	Watermelon
727.161.602.370.5	*ice-cream-white-seededii*	USA (MA)	Ice Cream White Seeded	Small (20lb) Red	Watermelon
727.161.602.370.6	*klondykei*	USA (CA)	Klondyke Blocky	Small (20lb) Red	Watermelon

Registration #	Variety	Native Location	Variety	Specimen	Genome
727.161.602.370.7	*klondykei-2*	USA (CA)	Klondyke #2 Blocky	Small (20lb) Red	Watermelon
727.161.602.370.8	*klondykei-peacock-improva*	USA (CA)	Klondyke Peacock Improved Blocky	Small (20lb) Red	Watermelon
727.161.602.370.9	*klondykei-R7*	USA (CA)	Klondyke Resistant #7	Small (20lb) Red	Watermelon
727.161.602.371.0	*legacyi*	USA (Texas)	Legacy	Small (20lb) Red	Watermelon
727.161.602.371.1	*louisiana-queenii*	USA (LA)	Louisiana Queen Blunt	Small (20lb) Red	Watermelon
727.161.602.371.2	*melitopoliskyi*	Asia (Russia)	Melitopolisky	Small (20lb) Red	Watermelon
727.161.602.371.3	*peacockii-WR60*	USA (CA)	Peacock Wilt Resistant #60	Small (20lb) Red	Watermelon
727.161.602.371.4	*small-shining-lightii*	Asia (Russia)	Shining Light	Small (20lb) Red	Watermelon
727.161.602.371.5	*sugarleei*	USA (TX)	Sugar Lee	Small (20lb) Red	Watermelon
727.161.602.371.6	*sunsweetii*	USA (PA)	Sun Sweet	Small (20lb) Red	Watermelon
727.161.602.371.7	*sunsugari*	USA (PA)	Sun Sugar	Small (20lb) Red	Watermelon
727.161.602.371.8	*winter-queenii*	USA (OR)	Winter Queen	Small (20lb) Red	Watermelon

727.161.602.38 **SPECIES 38** *lanatus-rubra-mini* Mini (10lb) Red Watermelon

Registration #	Variety	Native Location	Variety	Specimen	Genome
727.161.602.380.1	*rubra-mini*	USA	Original	Mini (10lb) Red	Watermelon
727.161.602.380.2	*cole's-earlyi*	USA	Cole's Early	Mini (10lb) Red	Watermelon
727.161.602.380.3	*charleston-grayi-bushii*	USA (SC)	Charleston Gray Bush	Mini (10lb) Red	Watermelon
727.161.602.380.4	*crimson-treatii*	USA (Ohio)	Crimson Treat Blocky	Mini (10lb) Red	Watermelon
727.161.602.380.5	*early-canadensis*	Canada (NF)	Early Canadal	Mini (10lb) Red	Watermelon
727.161.602.380.6	*early-midget*	Canada (NF)	Early Canada	Mini (10lb) Red	Watermelon
727.161.602.380.7	*golden-midgetii*	USA (AR)	Golden Midget	Mini (10lb) Red	Watermelon
727.161.602.380.8	*icebox-midgetii*	USA (ID)	Icebox Midget Round	Mini (10lb) Red	Watermelon
727.161.602.380.9	*jubileei-bushii*	USA (OH)	Jubilee Bush	Mini (10lb) Red	Watermelon
727.161.602.381.0	*king-winteri*	USA (CA)	King Winter	Mini (10lb) Red	Watermelon
727.161.602.381.1	*micky-leei*	USA (FL)	Micky Lee	Mini (10lb) Red	Watermelon
727.161.602.381.2	*navajo-sweetii*	USA (NM)	Navaho Sweet Round	Mini (10lb) Red	Watermelon
727.161.602.381.3	*navajo-winteri*	USA (AZ)	Navaho Winter	Mini (10lb) Red	Watermelon
727.161.602.381.4	*scaley-barkii*	USA (SC)	Scaley Bark	Mini (10lb) Red	Watermelon
727.161.602.381.5	*scaley-barkii-whitei*	USA (SC)	Scaley Bark White	Mini (10lb) Red	Watermelon
727.161.602.381.6	*sugarcreekii*	USA (PA)	Sugarcreek	Mini (10lb) Red	Watermelon
727.161.602.381.7	*super-sweetii*	USA (KS)	Super Sweet	Mini (10lb) Red	Watermelon
727.161.602.381.8	*sweet-favoritei*	Africa (Namibia)	Sweet Favorite	Mini (10lb) Red	Watermelon
727.161.602.381.9	*winter-kingii*	USA (SC)	Winter King	Mini (10lb) Red	Watermelon

727.161.602.39 **SPECIES 39** *lanatus-rubra-mini* Mini (10lb) Red Watermelon

Registration #	Variety	Native Location	Variety	Specimen	Genome
727.161.602.390.1	*rubra-micro*	USA	Original	Tiny (5-8lb) Red	Watermelon
727.161.602.390.2	*far-northii*	USA (MT)	Far North	Tiny (5-8lb) Red	Watermelon
727.161.602.390.3	*fourth-of-julyi*	USA (ID)	Fourth Of July	Tiny (5-8lb) Red	Watermelon
727.161.602.390.4	*garden-babyi*	USA (IL)	Garden Baby Round	Tiny (5-8lb) Red	Watermelon
727.161.602.390.5	*hopii-redii*	USA (NM)	Hopi Round	Tiny (5-8lb) Red	Watermelon
727.161.602.390.6	*ice-creamii-black-seededii*	USA (VA)	Ice Cream Black Seeded Round	Tiny (5-8lb) Red	Watermelon
727.161.602.390.7	*minileei*	USA (CO)	Mini Lee	Tiny (5-8lb) Red	Watermelon
727.161.602.390.8	*new-hampshire-midgetii*	USA (NH)	New Hampshire Midget	Tiny (5-8lb) Red	Watermelon
727.161.602.390.9	*red-cloudii*	Europe (Spain)	Red Cloud	Tiny (5-8lb) Red	Watermelon
727.161.602.391.0	*russica*	Asia (Russia)	Russian	Tiny (5-8lb) Red	Watermelon
727.161.602.391.1	*yugoslavianii*	W. Asia (Serbia)	Yugoslavian	Tiny (5-8lb) Red	Watermelon

The Complete Book of Fruits, ISBN:0670-86752-7, page, 114, states that "Melons are characterized by a thick, inedible rind covering succulent melting flesh, which encloses a central cavity and a battalion of flat, pointed oval whitish seeds. The first reference to really delicious aromatic melons comes in the fifteenth and sixteenth centuries probably as a result of hybridizations between many different strains". In the listings below, smooth skinned melons are referred to as Casaba or Honeydew. Orange flesh netted skin melons are referred to as Cantaloupe. Other netted melons are referred to as Muskmelons.

GENUS 03 *Cucumis-melo* Melon

CASABA, WHITE (*Cucumis-melo casaba-alba*)

727.161.603.01 **SPECIES 01** *casaba-alba* White Casaba Melon

Registration #	Variety	Native Location	Variety	Specimen	Genome
727.161.603.010.1	*alba*	M. East (Kassabeh, Turkey)	Original	White Casaba	Melon
727.161.603.010.2	*canariensis*	Pac. Is. (Canary Is.)	Canary 6lb	White Casaba	Melon

Registration #	Variety	Native Location	Variety	Specimen	Genome
727.161.603.010.3	*golden-beautyi*	USA (AZ, CA)	Golden Beauty 7.5lb	White Casaba	Melon
727.161.603.010.4	*marygoldii*	USA (Beltsville, MD)	Marygold 3.5lb	White Casaba	Melon
727.161.603.010.5	*ribertejo-blancoi*	Europe (Spain)	Ribertejo Blanco	White Casaba	Melon

CASABA, ORANGE (*Cucumis-melo casaba-aurantia*)

727.161.603.02 **SPECIES 02** *casaba-aurantia* Orange Casaba Melon

Registration #	Variety	Native Location	Variety	Specimen	Genome
727.161.603.020.1	*aurantia*	M. East (Kassabeh, Turkey)	Original	Orange Casaba	Melon
727.161.603.020.2	*bidwellii*	M. East (Kassabeh, Turkey)	Bidwell 14lb	Orange Casaba	Melon
727.161.603.020.3	*santa-domingoensis*	USA (New Mexico)	Santa Domingo	Orange Casaba	Melon

CASABA, GREEN (*Cucumis-melo casaba-groena*)

727.161.603.03 **SPECIES 03** *casaba-groena* Green Casaba Melon

Registration #	Variety	Native Location	Variety	Specimen	Genome
727.161.603.030.1	*groena*	M. East (Kassabeh, Turkey)	Original	Green Casaba	Melon
727.161.603.030.2	*santa-clausi*	USA	Santa Claus 8lb	Green Casaba	Melon
727.161.603.030.3	*santa-domingoensis*	USA (New Mexico)	Santa Domingo	Green Casaba	Melon
727.161.603.030.4	*sungoldii*	USA (New Hampshire)	Sungold	Green Casaba	Melon
727.161.603.030.5	*oodham-keli-basoi*	USA (Arizona)	Oodham Keli Baso	Green Casaba	Melon
727.161.603.030.6	*tam-mayan-sweetii*	USA (Texas)	Tam Mayan Sweet	Green Casaba	Melon

CHITO MELON (*Cucumis-melo chito*)

727.161.603.04 **SPECIES 04** *chito* Chito Melon

Registration #	Variety	Native Location	Variety	Specimen	Genome
727.161.603.040.1	*chito*	W. Indies	Original	Chito	Melon
727.161.603.040.2	*mango-melonii*	USA (Idaho)	White Mango 3"	Chito	Melon
727.161.603.040.3	*vine-peachii*	USA	Orange Yellow Peach 3"	Chito	Melon

PICKLING MELON (*Cucumis-melo conomon*)

727.161.603.05 **SPECIES 05** *conomon* Pickling Melon

Registration #	Variety	Native Location	Variety	Specimen	Genome
727.161.603.050.1	*conomon*	Europe, Asia	Original	Pickling	Melon
727.161.603.050.2	*ao-uri*	E.Asia (Japan)	Ao Uri White	Pickling	Melon
727.161.603.050.3	*half-long-of-puglia*	Europe (Puglia, Italy)	Half Long Of Puglia	Pickling	Melon
727.161.603.050.4	*italian-cucumber-fuzzy-whitei*	Europe (Italy)	Italian Cucumber Fuzzy White	Pickling	Melon
727.161.603.050.5	*katsura*	E. Asia (Japan)	Katsura 4" x 14"	Pickling	Melon
727.161.603.050.6	*kin-makuwai*	E. Asia (Japan)	Kin Makuwa	Pickling	Melon
727.161.603.050.7	*oriental-pickling-melonii*	E. Asia (Japan)	Oriental 3" x 12"	Pickling	Melon
727.161.603.050.8	*oshiro-uri*	E. Asia (Japan)	Oshiro Uri	Pickling	Melon
727.161.603.050.9	*numame-pickling-melonii*	E. Asia (Japan)	Numame 11"	Pickling	Melon

RIBBED MELON (*Cucumis-melo costata*)

727.161.603.06 **SPECIES 06** *costata* Ribbed Melon

Registration #	Variety	Native Location	Variety	Specimen	Genome
727.161.603.060.1	*costata*	Europe	Original	Ribbed	Melon
727.161.603.060.2	*isleta-puebloio*	USA (Albuquerque, NM)	Isleta Pueblo	Ribbed	Melon
727.161.603.060.3	*new-mexico-melonii*	USA (Alameda, NM)	New Mexico	Ribbed	Melon
727.161.603.060.4	*tortarello-abruozzesei*	Europe (Abruzzo, Italy)	Tortarello Abruzzo	Ribbed	Melon

POMEGRANATE MELON (*Cucumis-melo dudaim*)

727.161.603.07 **SPECIES 07** *dudaim* Pomegranate Melon

Registration #	Variety	Native Location	Variety	Specimen	Genome
727.161.603.070.1	*dudaim*	W. Indies	Original	Pomegranate	Melon
727.161.603.070.2	*plum-granniei*	USA (Tennessee)	Plum Grannie	Pomegranate	Melon

SNAKE MELON (*Cucumis-melo flexuosus*)

727.161.603.08 **SPECIES 08** *flexuosus* Snake Melon

Registration #	Variety	Native Location	Variety	Specimen	Genome
727.161.603.080.1	*flexuosus*	Europe, Asia	Original	Snake	Melon
727.161.603.080.2	*armeniaca*	Europe (Armenia)	Armenian 30"	Snake	Melon
727.161.603.080.3	*chinensis-longa*	Asia (China)	China Long 30"	Snake	Melon
727.161.603.080.4	*kyoto-three-feetii*	E. Asia (Japan)	Kyoto Three Feet 30"	Snake	Melon
727.161.603.080.5	*metki-dark-greenii*	Europe (France)	Metki Dark Green 18"	Snake	Melon

Registration #	Variety	Native Location	Variety	Specimen	Genome
727.161.603.080.6	*metki-painted-serpent*	Europe (France)	Metki Painted Serpent 18"	Snake	Melon
727.161.603.080.7	*metki-whitei*	Europe (France)	Metki White 18"	Snake	Melon
727.161.603.080.8	*serpent-cucumberi*	Europe (France)	Serpent Cucumber 50"	Snake	Melon
727.161.603.080.9	*white-striped-snakei*	Europe (France)	White Striped 30"	Snake	Melon

HONEYDEW MELON, WHITE (*Cucumis-melo inodorus-alba*)

727.161.603.09 **SPECIES 09** *inodorus-alba* White Honeydew Melon

Registration #	Variety	Native Location	Variety	Specimen	Genome
727.161.603.090.1	*alba*	W. Indies	Original	White Honeydew	Melon
727.161.603.090.2	*amarillo-oroi*	Europe (Spain)	Amarillo Oro 15lb	White Honeydew	Melon
727.161.603.090.3	*canariensis-96*	Europe (Italy)	Canary 1996	White Honeydew	Melon
727.161.603.090.4	*charlynnei*	Europe	Charlynne 5.5lb	White Honeydew	Melon
727.161.603.090.5	*cob-melonii*	USA (Arkansas)	Cob Melon	White Honeydew	Melon
727.161.603.090.6	*crenshaw-bella-dulcea*	Europe (Italy)	Crenshaw Bella Dulce	White Honeydew	Melon
727.161.603.090.7	*d'algiersi*	Europe (France)	D'Algiers	White Honeydew	Melon
727.161.603.090.8	*early-silverlinei*	USA (Decorah, IA)	Early Silverline1.5lb	White Honeydew	Melon
727.161.603.090.9	*giallo-da-invernoi-3*	Europe (Italy)	Gaillo Da Inverno #3 4.5lb	White Honeydew	Melon
727.161.603.091.0	*golden-crispyi*	E. Asia (Japan)	Golden Crispy	White Honeydew	Melon
727.161.603.091.1	*honey-goldii-9*	E. Asia (Japan)	Honey Gold #9 10oz	White Honeydew	Melon
727.161.603.091.2	*kazakhensis*	E. Europe (Kazakstan)	Kazakh 2lb	White Honeydew	Melon
727.161.603.091.3	*orientalis-golden-sweetii*	E. Asia (Japan)	Oriental Golden Sweet 3"	White Honeydew	Melon
727.161.603.091.4	*orientalis-jade-kingii*	E. Asia (Japan)	Oriental Jade King	White Honeydew	Melon
727.161.603.091.5	*pinyonet-de-valencia*	Europe (Spain)	Pinyonet De Valencia	White Honeydew	Melon
727.161.603.091.6	*swan-lakei*	USA (CA, NM)	Swan Lake 2lb	White Honeydew	Melon
727.161.603.091.7	*tiggeri*	Europe (Armenia)	Tigger 1lb	White Honeydew	Melon
727.161.603.091.8	*valenciensis*	Europe (Spain)	Valencia	White Honeydew	Melon
727.161.603.091.9	*valenciensis-praecox*	Europe (Spain)	Valencia Early	White Honeydew	Melon
727.161.603.092.0	*valenciensis-espanholii*	Europe (Spain)	Valencia Espanhol	White Honeydew	Melon
727.161.603.092.1	*valenciensis-tarda*	Europe (Spain)	Valencia Late	White Honeydew	Melon

HONEYDEW MELON, ORANGE (*Cucumis-melo inodorus-aurantia*)

727.161.603.10 **SPECIES 10** *inodorus-aurantia* Orange Honeydew Melon

Registration #	Variety	Native Location	Variety	Specimen	Genome
727.161.603.100.1	*aurantia*	Europe	Original	Orange Honeydew	Melon
727.161.603.100.2	*early-black-rockii*	Europe (France)	Early Black Rock 3lb	Orange Honeydew	Melon
727.161.603.100.3	*kroumiri*	Europe (France)	Kroumir 2lb	Orange Honeydew	Melon
727.161.603.100.4	*navajo-mixta-aurantia*	USA (NM)	Navaho Mix	Orange Honeydew	Melon
727.161.603.100.5	*noir-des-carmesi*	Europe (France)	Noir Des Carmes 2.5lb	Orange Honeydew	Melon
727.161.603.100.6	*orange-fleshed-honeydewii*	USA	Creamy Skin 5lb	Orange Honeydew	Melon
727.161.603.100.7	*printadoui*	Europe (France)	Printadou	Orange Honeydew	Melon
727.161.603.100.8	*savori*	Europe (France)	Savor 2.25lb	Orange Honeydew	Melon
727.161.603.100.9	*stutz-supremei*	USA (Chico, CA)	Stutz Supreme 8lb	Orange Honeydew	Melon
727.161.603.101.0	*sweet-frecklesii*	USA (Oregon)	Sweet Freckles	Orange Honeydew	Melon
727.161.603.101.1	*vedrantias*	Europe (France)	Vedrantias 1.5lb	Orange Honeydew	Melon

HONEYDEW MELON, SALMON (*Cucumis-melo inodorus-aurantia-rubra*)

727.161.603.11 **SPECIES 11** *inodorus-aurantia-rubra* Salmon Honeydew Melon

Registration #	Variety	Native Location	Variety	Specimen	Genome
727.161.603.110.1	*aurantia-rubra*	Europe	Original	Salmon Honeydew	Melon
727.161.603.110.2	*charentias*	Europe (France)	Charentias 2lb	Salmon Honeydew	Melon
727.161.603.110.3	*crenshawii*	Europe (Italy)	Crenshaw 8lb	Salmon Honeydew	Melon
727.161.603.110.4	*crenshawii-blancoi*	Europe (Italy)	Crenshaw Blanco 5lb	Salmon Honeydew	Melon
727.161.603.110.5	*crenshawii-cranei*	USA (Santa Rosa,CA)	Crane	Salmon Honeydew	Melon
727.161.603.110.6	*honeyloupei*	USA (California)	Honeyloupe 4lb	Salmon Honeydew	Melon
727.161.603.110.7	*prescott-fond-blanci*	Europe (France)	Prescott Fond Blanc 7lb	Salmon Honeydew	Melon
727.161.603.110.8	*tigeri*	M. East (Afghanistan)	Tiger	Salmon Honeydew	Melon

HONEYDEW MELON, GREEN (*Cucumis-melo inodorus-groena*)

727.161.603.12 **SPECIES 12** *inodorus-groena* Green Honeydew Melon

Registration #	Variety	Native Location	Variety	Specimen	Genome
727.161.603.120.1	*groena*	E. Asia	Original 6lb	Green Honeydew	Melon
727.161.603.120.2	*jaune-des-canaries-1*	Europe (France)	Canary Yellow 5lb	Green Honeydew	Melon
727.161.603.120.3	*early-honeydewii*	Canada(Manitoba)	Early 2.5lb	Green Honeydew	Melon
727.161.603.120.4	*golden-honeymoonii*	USA (Washington)	Golden Honeymoon	Green Honeydew	Melon
727.161.603.120.5	*gold-rindii*	USA (Washington)	Gold Rind 5lb	Green Honeydew	Melon
727.161.603.120.6	*green-climbingii*	Europe (France)	Green Climbing 1lb	Green Honeydew	Melon
727.161.603.120.7	*navajo-mixta-groena*	USA (Shiprock, NM)	Navaho Mix	Green Honeydew	Melon
727.161.603.120.8	*ojo-calientei*	USA (New Mexico)	Ojo Caliente 6lb	Green Honeydew	Melon
727.161.603.120.9	*piel-de-sapoi*	Europe (Spain)	Piel De Sapo 8lb	Green Honeydew	Melon
727.161.603.121.0	*sakata's-sweetii*	Europe	Sakata's Sweet 11oz	Green Honeydew	Melon
727.161.603.121.1	*san-juanensis*	USA (New Mexico)	San Juan	Green Honeydew	Melon
727.161.603.121.2	*sweet-delightii*	USA (Colorado)	Sweet Delight 6lb	Green Honeydew	Melon
727.161.603.121.3	*tam-dewii*	USA (Texas)	Tam Dew 5.5lb	Green Honeydew	Melon
727.161.603.121.4	*tam-dewii-improva*	USA (Texas)	Tam Dew Improved 5lb	Green Honeydew	Melon
727.161.603.121.5	*thai-round-greenii*	S. Asia (Thailand)	Thai Round	Green Honeydew	Melon

HONEYDEW MELON, YELLOW (*Cucumis-melo inodorus-helva*)

727.161.603.13 **SPECIES 13** *inodorus-helva* Yellow Honeydew Melon

Registration #	Variety	Native Location	Variety	Specimen	Genome
727.161.603.130.1	*helva*	S. Asia	Original	Yellow Honeydew	Melon
727.161.603.130.2	*escondido-goldii*	S. Asia (Tibet)	Escondido Gold 23lb	Yellow Honeydew	Melon
727.161.603.130.3	*melon-de-castilloi*	Mexico (Sierra Madre)	Melon De Castillo	Yellow Honeydew	Melon
727.161.603.130.4	*quitoi*	Europe (France)	Quito	Yellow Honeydew	Melon

727.161.603.14 **SPECIES 14** *momordica* Snap Melon

Registration #	Variety	Native Location	Variety	Specimen	Genome
727.161.603.140.1	*momordica*	S. Asia	Original	Snap	Melon
727.161.603.140.2	*kachrai*	S. Asia (India)	Kachra	Snap	Melon

727.161.603.15 **SPECIES 15** *metuliferus* Horned Jelly Melon

727.161.603.16 **SPECIES 16** *mollis* Soft Melon

Registration #	Variety	Native Location	Variety	Specimen	Genome
727.161.603.160.1	*mollis*	USA	Original	Soft	Melon
727.161.603.160.2	*million-dollari*	USA (Maine)	Million Dollar	Soft	Melon
727.161.603.160.3	*musa*	USA	Banana 7lb 20"	Soft	Melon
727.161.603.160.4	*souhela*	USA (California)	Souhela	Soft	Melon
727.161.603.160.5	*tardif-espagnolii*	Europe (Spain)	Tardif Espagnol	Soft	Melon

MUSKMELON, WHITE (*Cucumis-melo reticulatus-alba*)

727.161.603.17 **SPECIES 17** *reticulatus-alba* White Muskmelon Melon

Registration #	Variety	Native Location	Variety	Specimen	Genome
727.161.603.170.1	*alba*	Europe	Original	White Muskmelon	Melon
727.161.603.170.2	*amarillo-lisoi*	USA (Texas)	Amarillo Liso	White Muskmelon	Melon
727.161.603.170.3	*collective-farm-womenii*	E. Europe (Ukraine)	Collective Farm Woman	White Muskmelon	Melon
727.161.603.170.4	*israelii*	E. Europe (Ukraine)	Israeli	White Muskmelon	Melon
727.161.603.170.5	*spanish-moonii*	Europe (Spain)	Spanish Moon	White Muskmelon	Melon
727.161.603.170.6	*tendral-verde-tardifii*	Europe (France)	Tendral Verde Tardif	White Muskmelon	Melon
727.161.603.170.7	*tendral-verde-tardioi*	Europe (France)	Tendral Verde Tardio	White Muskmelon	Melon
727.161.603.170.8	*unclei-E*	USA (Washington)	Uncle E	White Muskmelon	Melon

CANTALOUPE (*Cucumis-melo reticulatus-aurantia*)

727.161.603.18 **SPECIES 18** *reticulatus-aurantia* Cantalope Melon

Registration #	Variety	Native Location	Variety	Specimen	Genome
727.161.603.180.1	*aurantia*	Europe (Cantaluppi Estate, Rome, Italy)	Original	Cantaloupe	Melon
727.161.603.180.2	*albersii*	Canada(Ontario)	Albers	Cantaloupe	Melon
727.161.603.180.3	*altai*	Asia (Altai, Mtns., Russia)	Altai Mountain	Cantaloupe	Melon
727.161.603.180.4	*amishii*	USA (Pennsylvania)	Amish 8lb	Cantaloupe	Melon
727.161.603.180.5	*ananas-d'amerique-a-chair-rougei*	Europe (France)	Pineapple Chair Rouge	Cantaloupe	Melon
727.161.603.180.6	*aurora*	USA (Alabama)	Aurora 4lb	Cantaloupe	Melon
727.161.603.180.7	*blenheimii*	UK (England)	Blenheim 1.5lb	Cantaloupe	Melon
727.161.603.180.8	*chimayoi*	Europe (Spain)	Chimayo	Cantaloupe	Melon
727.161.603.180.9	*early-large-prescottii*	Europe (France)	Early Large Prescott [illegible]lb	Cantaloupe	Melon
727.161.603.181.0	*edisto*	USA (S. Carolina)	Edisto 4lb	Cantaloupe	Melon
727.161.603.181.1	*eel-riveri*	E. Asia (Japan)	Eel River 7lb	Cantaloupe	Melon

Registration #	Variety	Native Location	Variety	Specimen	Genome
727.161.603.181.2	*emerald-greenii*	USA (Pennsylvania)	Emerald Green 2lb	Cantaloupe	Melon
727.161.603.181.3	*giant-perfectionii*	USA (S. Carolina)	Giant Perfection 16lb	Cantaloupe	Melon
727.161.603.181.4	*golden-gopheri*	USA (Maine)	Golden Gopher 3.5lb	Cantaloupe	Melon
727.161.603.181.5	*golden-jennyi*	USA (Virginia)	Golden Jenny 3/4lb	Cantaloupe	Melon
727.161.603.181.6	*granite-statei*	USA (Iowa)	Granite State 4lb	Cantaloupe	Melon
727.161.603.181.7	*gulfstreamii*	USA (S. Carolina)	Gulfstream 5.5"	Cantaloupe	Melon
727.161.603.181.8	*harvest-queenii*	USA (Wisconsin)	Harvest Queen 3.5lb	Cantaloupe	Melon
727.161.603.181.9	*hearts-of-goldii*	USA (Benton Harbor, MI)	Hearts Of Gold 4lb	Cantaloupe	Melon
727.161.603.182.0	*hearts-of-goldii-morrillii*	USA (Benton Harbor, MI)	Hearts Of Gold Morrill	Cantaloupe	Melon
727.161.603.182.1	*honey-combedii*	Canada (Ontario)	Honey Combed 6lb	Cantaloupe	Melon
727.161.603.182.2	*imperialis*	USA (New Mexico)	Imperial 4lb	Cantaloupe	Melon
727.161.603.182.3	*imperialis-45*	USA (Dixie)	Imperial #45 3.5lb	Cantaloupe	Melon
727.161.603.182.4	*imperialis-45-ECS*	USA (Dixie)	Imperial #45-ECS 3.5lb	Cantaloupe	Melon
727.161.603.182.5	*iroquoisi*	USA (New York)	Iroquois 6lb	Cantaloupe	Melon
727.161.603.182.6	*itsy-bitsy-sweetheartii*	USA (Ohio)	Itsy Bitsy Sweetheart	Cantaloupe	Melon
727.161.603.182.7	*kansasensis*	USA (Kansas)	Kansas 4lb	Cantaloupe	Melon
727.161.603.182.8	*old-west-virginica*	USA (W. Virginia)	Old West Virginia	Cantaloupe	Melon
727.161.603.182.9	*prescott-large-whitei*	Europe (France)	Prescott Large White 7lb	Cantaloupe	Melon
727.161.603.183.0	*magdalena*	Mexico	Magdalena	Cantaloupe	Melon
727.161.603.183.1	*mainstreamii*	USA	Mainstream 2.75lb	Cantaloupe	Melon
727.161.603.183.2	*market-pridei*	USA (S. Carolina)	Market Pride 4lb	Cantaloupe	Melon
727.161.603.183.3	*melba*	Europe (Poland)	Melba 1.25lb	Cantaloupe	Melon
727.161.603.183.4	*mountain-magici*	USA (New Mexico)	Mountain Magic 1lb	Cantaloupe	Melon
727.161.603.183.5	*north-carolinensis*	USA (N. Carolina)	North Carolina 25lb	Cantaloupe	Melon
727.161.603.183.6	*nutmegii*	USA	Nutmeg	Cantaloupe	Melon
727.161.603.183.7	*nutmegii-extra-earlyi*	USA	Nutmeg Extra Early	Cantaloupe	Melon
727.161.603.183.8	*okai*	Canada (Quebec)	Oka	Cantaloupe	Melon
727.161.603.183.9	*oregon-deliciosa*	USA (Oregon)	Oregon Delicious	Cantaloupe	Melon
727.161.603.184.0	*peari*	USA (N. Dakota)	Pear	Cantaloupe	Melon
727.161.603.184.1	*persica*	M. East (Iran)	Persian 6lb	Cantaloupe	Melon
727.161.603.184.2	*persica-jharbezeh-mashadii*	M. East(Iran)	Jharbezeh Mashadi Persian	Cantaloupe	Melon
727.161.603.184.3	*pikei*	USA (Virginia)	Pike 7lb	Cantaloupe	Melon
727.161.603.184.4	*oran's-melonii*	USA (Arkansas)	Oran's 7lb	Cantaloupe	Melon
727.161.603.184.5	*planters-jumboi*	USA (S. Carolina)	Planters Jumbo 4.25lb	Cantaloupe	Melon
727.161.603.184.6	*petit-gris-de-rennesi*	Europe (France)	Petit Gris De Rennes 2.5lb	Cantaloupe	Melon
727.161.603.184.7	*pride-of-wisconsinii*	USA (Wisconsin)	Pride Of Wisconsin 7"	Cantaloupe	Melon
727.161.603.184.8	*retato-degli-ortolanii*	Europe (Italy)	Retato Degli Ortolani 3.5lb	Cantaloupe	Melon
727.161.603.184.9	*sleeping-beautyi*	USA (Virginia)	Sleeping Beauty 1/2lb	Cantaloupe	Melon
727.161.603.185.0	*smith's-perfectii*	USA (Maryland)	Smith's Perfect 4lb	Cantaloupe	Melon
727.161.603.185.1	*speari*	USA (Oregon)	Spear 5lb	Cantaloupe	Melon
727.161.603.185.2	*sugar-rockii*	USA (Hicksville,OH)	Sugar Rock 4lb	Cantaloupe	Melon
727.161.603.185.3	*sweet-granitei*	USA (New Hampshire)	Sweet Granite 3lb	Cantaloupe	Melon
727.161.603.185.4	*sweet-passionii*	USA (Pennsylvania)	Sweet Passion 3.5lb	Cantaloupe	Melon
727.161.603.185.5	*tam-uvaldei*	USA (Texas)	Tam Uvalde 4lb	Cantaloupe	Melon
727.161.603.185.6	*thai-longa*	S. Asia (Thailand)	Thai Long	Cantaloupe	Melon
727.161.603.185.7	*thayeri*	USA (Kansas)	Thayer 10lb	Cantaloupe	Melon
727.161.603.185.8	*weeks-north-carolina-giantii*	USA (N. Carolina)	Weeks North Carolina Giant 22lb	Cantaloupe	Melon
727.161.603.185.9	*vandehay-familyi*	USA (Washington)	Vandahay Family	Cantaloupe	Melon
727.161.603.185.9	*zattai*	USA (Italy)	Zatta 4lb	Cantaloupe	Melon

MUSKMELON, ORANGE YELLOW (*Cucumis-melo reticulatus-aurantia-helva*)

727.161.603.19 **SPECIES 19** *reticulatus-aurantia-helva* Orange Yellow Muskmelon Melon

Registration #	Variety	Native Location	Variety	Specimen	Genome
727.161.603.190.1	*aurantia-helva*	Europe	Original	Orange Yellow Muskmelon	Melon
727.161.603.190.2	*castella*	Europe	Castella 3lb	Orange Yellow Muskmelon	Melon
727.161.603.190.3	*deliciosa*	USA (New York)	Delicious 3.5lb	Orange Yellow Muskmelon	Melon
727.161.603.190.4	*emerald-gemii*	USA (Pennsylvania)	Emerald Gem 3lb	Orange Yellow Muskmelon	Melon
727.161.603.190.5	*ginger's-pridei*	USA (Kentucky)	Ginger's Pride	Orange Yellow Muskmelon	Melon
727.161.603.190.6	*jake'si*	USA (New Mexico)	Jake's	Orange Yellow Muskmelon	Melon
727.161.603.190.7	*navajo-yellowii*	USA (New Mexico)	Navaho	Orange Yellow Muskmelon	Melon

MUSKMELON, SALMON (*Cucumis-melo reticulatus-aurantia-rubra*)

727.161.603.20 **SPECIES 20** *reticulatus-aurantia-rubra* Salmon Muskmelon Melon

Registration #	Variety	Native Location	Variety	Specimen	Genome
727.161.603.200.1	*aurantia-rubra*	USA	Original	Salmon Muskmelon	Melon
727.161.603.200.2	*bender's-surprisei*	USA (Ohio)	BenderSurprise 7lb	Salmon Muskmelon	Melon
727.161.603.200.3	*boughemii*	Canada (Saskatoon)	Boughem	Salmon Muskmelon	Melon
727.161.603.200.4	*burrell's-jumboi*	USA (Colorado)	Burrell's Jumbo 5lb	Salmon Muskmelon	Melon
727.161.603.200.5	*deliciosa-51*	USA (New York)	Delicious #51 4.5lb	Salmon Muskmelon	Melon
727.161.603.200.6	*deliciosa-51earli*	USA (New York)	Delicious #51 Early 5.5lb	Salmon Muskmelon	Melon
727.161.603.200.7	*deliciosa-goldenii*	USA (New York)	Delicious Golden	Salmon Muskmelon	Melon
727.161.603.200.8	*dr.-jaegar'si*	USA (S. Carolina)	Dr. Jeagar's	Salmon Muskmelon	Melon
727.161.603.200.9	*early-hanoveri*	USA (Pennsylvania)	Early Hanover	Salmon Muskmelon	Melon
727.161.603.201.0	*edisto-47*	USA (S. Carolina)	Edisto #47 4.5lb	Salmon Muskmelon	Melon
727.161.603.201.1	*far-northii*	Canada (Saskatchewan)	Far North	Salmon Muskmelon	Melon
727.161.603.201.2	*hale's-bestii*	USA (Dixie)	Hale's Best 4lb	Salmon Muskmelon	Melon
727.161.603.201.3	*hale's-bestii-36*	USA (Dixie)	Hale's Best #36 3.25lb	Salmon Muskmelon	Melon
727.161.603.201.4	*hale's-bestii-4-50*	USA (Texas)	Hale's Best #4-50 3.5lb	Salmon Muskmelon	Melon
727.161.603.201.5	*hale's-bestii-936*	USA (Dixie)	Hale's Best #936 4lb	Salmon Muskmelon	Melon
727.161.603.201.6	*hearts-of-goldii-improva*	USA (Benton Harbor, MI)	Hearts Of Gold Improved 3.5lb	Salmon Muskmelon	Melon
727.161.603.201.7	*hollybrook-lusciousi*	USA (Richmond,VA)	Hollybrook Luscious 9lb	Salmon Muskmelon	Melon
727.161.603.201.8	*honeybushii*	USA (Oregon)	Honeybush 2.75lb	Salmon Muskmelon	Melon
727.161.603.201.9	*honey-rockii*	USA (Michigan)	Honey Rock 3.5lb	Salmon Muskmelon	Melon
727.161.603.202.0	*kangoldii*	USA (Kansas)	Kangold 4.5lb	Salmon Muskmelon	Melon
727.161.603.202.1	*musketeeri*	USA	Musketeer 4lb	Salmon Muskmelon	Melon
727.161.603.202.2	*northern-arizonensis*	USA (Arizona)	Northern Arizona 5"	Salmon Muskmelon	Melon
727.161.603.202.3	*perlita*	USA (Texas)	Perlita 3lb	Salmon Muskmelon	Melon
727.161.603.202.4	*rampicante-zuccherinoi*	Europe (Italy)	Rampicante Zuccherino 2lb	Salmon Muskmelon	Melon
727.161.603.202.5	*rocky-fordii*	USA (Colorado)	Rocky Ford 2.5lb	Salmon Muskmelon	Melon
727.161.603.202.6	*schoon-hard-shellii*	USA (Illinois)	Schoon Hard Shell 7lb	Salmon Muskmelon	Melon
727.161.603.202.7	*shumway's-giantii*	USA (S. Carolina)	Shumway's Giant 9lb	Salmon Muskmelon	Melon
727.161.603.202.8	*sierra-goldii*	USA	Sierra Gold 9lb	Salmon Muskmelon	Melon
727.161.603.202.9	*spartan-rockii*	USA (Michigan)	Spartan Rock 5"	Salmon Muskmelon	Melon
727.161.603.203.0	*sweetheartii*	USA (Vermont, 1994)	Sweetheart 4"	Salmon Muskmelon	Melon
727.161.603.203.1	*topmarkii*	USA (Colorado)	Top Mark 6.5"	Salmon Muskmelon	Melon
727.161.603.203.2	*topmarkii-crownsetii*	USA (Colorado)	Top Mark Crownset	Salmon Muskmelon	Melon
727.161.603.203.3	*top-netii-SR*	USA (California)	Top Net SR	Salmon Muskmelon	Melon
727.161.603.203.4	*turkeyi*	USA (Iowa)	Turkey 13lb	Salmon Muskmelon	Melon

MUSKMELON, GREEN (*Cucumis-melo reticulatus-groena*)

727.161.603.22 **SPECIES 21** *reticulatus-groena* Green Muskmelon Melon

Registration #	Variety	Native Location	Variety	Specimen	Genome
727.161.603.210.1	*groena*	E. Asia	Original	Green Muskmelon	Melon
727.161.603.210.2	*amuri*	E. Asia (Japan)	Amur	Green Muskmelon	Melon
727.161.603.210.3	*ananas-d'amerique-a-chair-vertei*	Europe (France)	Pineapple Chair Verte	Green Muskmelon	Melon
727.161.603.210.4	*anne-arundelii*	USA (Anne Arundel, MD)	Anne Arundel 3.5lb	Green Muskmelon	Melon
727.161.603.210.5	*boule-d'ori*	Europe (France)	Boule D'Or	Green Muskmelon	Melon
727.161.603.210.6	*bush-midgetii*	USA (Missouri)	Bush Midget	Green Muskmelon	Melon
727.161.603.210.7	*cavaillon-espagnola*	Europe (Spain)	Cavaillon Espagnol 6lb	Green Muskmelon	Melon
727.161.603.210.8	*green-machinei*	USA (Virginia)	Green Machine	Green Muskmelon	Melon
727.161.603.210.9	*green-nutmegii*	USA	Green Nutmeg 2.5lb	Green Muskmelon	Melon
727.161.603.211.0	*green-bushii*	USA (Missouri)	Green Bush 3"	Green Muskmelon	Melon
727.161.603.211.1	*ha-ogenii*	M. East (Israel)	Israel Ha-Ogen 4lb	Green Muskmelon	Melon
727.161.603.211.2	*jenny-lindii*	E. Asia (Armenia)	Jenny Lind 2lb	Green Muskmelon	Melon
727.161.603.211.3	*montrealensis*	Canada (Montreal)	Montreal 20lb	Green Muskmelon	Melon
727.161.603.211.4	*rocky-fordii*	USA (Colorado)	Rocky Ford 2.5lb	Green Muskmelon	Melon

MUSKMELON, YELLOW (*Cucumis-melo reticulatus-helva*)

727.161.603.22 **SPECIES 22** *reticulatus-helva* Ribbed Muskmelon Melon

Registration #	Variety	Native Location	Variety	Specimen	Genome
727.161.603.220.1	*helva*	M. East	Original	Yellow Muskmelon	Melon
727.161.603.220.2	*ananas*	M. East (Israel)	Pineapple 7lb	Yellow Muskmelon	Melon

Registration #	Variety	Native Location	Variety	Specimen	Genome
727.161.603.220.3	*champlainii*	USA (New York)	Golden Champlain	Yellow Muskmelon	Melon
727.161.603.220.4	*goldmasteri*	USA (California)	Goldmaster	Yellow Muskmelon	Melon
727.161.603.220.5	*minnesota-midgetii*	USA (Minnesota)	Minnesota Midget	Yellow Muskmelon	Melon
727.161.603.220.6	*montana-goldii*	USA (Belgrade, MT)	Montana Gold	Yellow Muskmelon	Melon
727.161.603.220.7	*sharlynii*	USA (Washington)	Sharlyn 5lb	Yellow Muskmelon	Melon

WRINKLED MELON (*Cucumis-melo rugosa*)

727.161.603.23 **SPECIES 23** *rugosa* Wrinkled Melon

Registration #	Variety	Native Location	Variety	Specimen	Genome
727.161.603.230.1	*rugosa*	Europe	Original	Wrinkled	Melon
727.161.603.230.2	*carosello-mezzo-lungo-baresei*	Europe (Bari, Italy)	Carosello Mezzo Lungo Barese	Wrinkled	Melon
727.161.603.230.3	*giallo-canary-rugosoi*	Europe (Italy)	Giallo Canary Rugoso	Wrinkled	Melon
727.161.603.230.4	*old-time-tennesseei*	USA (Tennessee)	Old Time Tennessee	Wrinkled	Melon
727.161.603.230.5	*spanishii*	Europe (Spain)	Spanish Green & Gold 10lb	Wrinkled	Melon
727.161.603.230.6	*tortarello-baresei*	Europe (Bari, Italy)	Tortarello Bari	Wrinkled	Melon

CUCUMBER MELON (*Cucumis-melo sativa*)

727.161.603.24 **SPECIES 24** *sativa* Cucumber Melon

Registration #	Variety	Native Location	Variety	Specimen	Genome
727.161.603.240.1	*sativa*	M. East	Original	Cucumber	Melon
727.161.603.240.2	*the-dukei*	The Duke	Original	Cucumber	Melon
727.161.603.240.3	*tondo-liscia-manduria*	Europe (Italy)	Tondo Liscia Manduria	Cucumber	Melon

GENUS *Cucubertia* (Synonym for *Citrullus*)

BALSAM PEAR (*Momordica*)

GENUS 04 *Momordica* Balsam Pear

727.161.604.01 **SPECIES 01** *balsamina* Bitter Melon Balsam Pear

Registration #	Variety	Native Location	Variety	Specimen	Genome
727.161.604.010.1	*balsamina*	S. Asia	Original	Bitter Melon	Balsam Pear
727.161.604.010.2	*hongkongensis*	S. Asia (Hong Kong)	Hong Kong	Bitter Melon	Balsam Pear
727.161.604.010.3	*gigantea*	S. Asia	Large	Bitter Melon	Balsam Pear
727.161.604.010.4	*thailandica*	S. Asia (Thailand)	Thailand	Bitter Melon	Balsam Pear

727.161.604.02 **SPECIES 02** *charantia* Bitter Gourd Balsam Pear

Registration #	Variety	Native Location	Variety	Specimen	Genome
727.161.604.020.1	*charantia*	.	Original	Bitter Gourd	Balsam Pear
727.161.604.020.2	*improva*	.	Improved	Bitter Gourd	Balsam Pear
727.161.604.020.3	*fontana*	.	Fontana	Bitter Gourd	Balsam Pear

727.161.604.03 **SPECIES 03** *cochinchinensis* Cochin China Balsam Pear

727.161.604.04 **SPECIES 04** *dioica* Dioecious Balsam Pear

CHAYOTE PLUM (*Sechium*)

Flora, ISBN:1-74048-097-X, page 1328, states that "The large, green, ridged, oval fruits are fleshy and have hairy, spiny or smooth skin. They contain a single seed that can germinate within the fruit". Flora, ISBN:1-74048-097-X, page 1328, states that for *Sechium edule*; Edible greenish yellow fruits to 7 in long".

GENUS 05 *Sechium* Chayote

727.161.605.01 **SPECIES 01** *edule* Christophine Chayote

CALEBASSE (*Sicana*)

GENUS 06 *Sicana* Calebasse

The Complete Book of Fruits, ISBN:0670-86752-7, page, 245, states that for *Sicana odorifera*; "A perennial [vine] in Brazil and Equador, . . . it has yellow flowers and yellowish red fruits with a sweet flavor. The fruits can be eaten young as vegetables and ripe as fruits, most commonly in jams".

727.161.606.01 **SPECIES 01** *odorifera* Casa Banana Calebasse

Governor's Plum Family (*Flacourtiaceae*)

Family 7 *Flacourtiaceae* Flacourt

GENUS *Aberia* (Synonym for *Dovyalis*)

GOVERNER'S PLUM (*Flacourtia*)

Flora, ISBN:1-74048-097-X, page 606, states that "Fruit is a round, smooth, berry-like drupe with 8 to 12 flat seeds in a sweet juicy pulp used in jams and perserves".

GENUS 01 *Flacourtia* Governor's Plum

727.161.701.01 **SPECIES 01** *indica* Batoko Governor's Plum

Registration #	Variety	Native Location	Variety	Specimen	Genome
727.161.701.010.1	*indica*	S. Asia (India), Africa (Zambia), Madagascar	Original	Batoko	Governor's Plum
727.161.701.010.2	*ramontchi*	S. Asia (India), Africa (Zambia), Madagascar	Ramontch	Batoko	Governor's Plum

727.161.701.02 **SPECIES 02** *inermis* Thornless Governor's Plum
727.161.701.03 **SPECIES 03** *jangomas* Paniala Governor's Plum

Registration #	Variety	Native Location	Variety	Specimen	Genome
727.161.701.030.1	*jangomas*	S. Asia (India), E. Australia	Original	Paniala	Governor's Plum
727.161.701.030.2	*cataphracta*	S. Asia (India), E. Australia	Cataphract	Paniala	Governor's Plum

727.161.701.04 **SPECIES 04** *rukam* Filimoto Governor's Plum

Registration #	Variety	Native Location	Variety	Specimen	Genome
727.161.701.040.1	*rukam*	S. Asia (Malaysia), Philippines	Original	Filimoto	Governor's Plum
727.161.701.040.2	*euphlebia*	S. Asia (Malaysia), Philippines	Phlebe	Filimoto	Governor's Plum

Flora, ISBN:1-74048-097-X, page 606, states that for *Dovyalis* "Fruits spherical, fleshy, sometimes used for pickles and jams".

GENUS *Aberia* (Synonym for *Dovyalis*)
GENUS 02 *Dovyalis* Dovy Fruit

KOSHUM PLUM (*Dovyalis abyssinica*)

727.161.702.01 **SPECIES 01** *abyssinica* Koshum Plum Dovy Fruit

KEI APPLE (*Dovyalis caffra*)

727.161.702.02 **SPECIES 02** *caffra* Kei Apple Dovy Fruit

Registration #	Variety	Native Location	Variety	Specimen	Genome
727.161.702.020.1	*caffra*	S. Africa (E. Cape, Mozambique, Malawi)	Original	Kei Apple	Dovy Fruit
727.161.702.020.2	*sweet-clonei*	S. Africa (E. Cape, Mozambique, Malawi)	Sweet Clone	Kei Apple	Dovy Fruit

CEYLON GOOSEBERRY (*Dovyalis hebecarpa*)

727.161.702.03 **SPECIES 03** *hebecarpa* Ceylon Gooseberry Dovy Fruit

KOSHUM GOOSEBERRY (*Dovyalis hebecarpa x abyssinica*)

727.161.702.04 **SPECIES 04** *hebecarpa x abyssinica* Koshum Gooseberry Dovy Fruit

DOVY APRICOT (*Dovyalis hebecarpa x abyssinica*)

727.161.702.05 **SPECIES 05** *hebecarpa x caffra* Tropical Apricot Dovy Fruit

Registration #	Variety	Native Location	Variety	Specimen	Genome
727.161.702.050.1	*caffra*	S. Asia (Sri Lanka)	Original	Tropical Apricot	Dovy Fruit
727.161.702.050.2	*prodigalis*	S. Asia (Sri Lanka)	Prodigal	Tropical Apricot	Dovy Fruit

LOGWOOD CHERRY (*Xylosma*)

GENUS 03 *Xylosma* Logwood Cherry
727.161.703.01 **SPECIES 01** *buxifolia* Mucha Gente Logwood Cherry

Fruit, Berry and Nut Inventory, 3rd Ed., ISBN:1-882424-56-5, states that for *Xylosma controversum*, "Small red fruit ripens from July to November".

727.161.703.02 **SPECIES 02** *controversum* Tulo Sauuji Logwood Cherry
727.161.703.03 **SPECIES 03** *crenata* Sawtooth Logwood Cherry
727.161.703.04 **SPECIES 04** *flexuosa* Brush Holly Logwood Cherry
727.161.703.05 **SPECIES 05** *hawaiiensis* Hawaiian Brush Holly Logwood Cherry

Registration #	Variety	Native Location	Variety	Specimen	Genome
727.161.703.050.1	*hawaiiensis*	Pac. Is. (Hawaiian Is.)	Original	Hawaiian Brush Holly	Logwood Cherry
727.161.703.050.2	*hillebrandii*	Pac. Is. (Hawaiian Is.)	Hillebrand	Hawaiian Brush Holly	Logwood Cherry

727.161.703.06 **SPECIES 06** *pachyphylla* Spiny Logwood Cherry
727.161.703.07 **SPECIES 07** *schaefferioides* White Logwood Cherry
727.161.703.08 **SPECIES 08** *schwaneckeana* Schwaneck Logwood Cherry

Caper Order (*Caperales*)

Order 2 *Caperales* Caper

Caper Family (*Caperaceae*)

Family 1 *Caperaceae* Caper

CAPER BERRY (*Capparis*)

GENUS 01 *Capparis* Caper

Flora, ISBN:1-74048-097-X, page 325, states that "The fruit is berry-like". Flora, ISBN:1-74048-097-X, page 326, states that for *Capparis arborea*; "Edible green fruit in autumn".

727.162.101.01 **SPECIES 01** *arborea* Berry Caper

727.162.101.02 **SPECIES 02** *amplissima* Burro Blanco Caper

Registration #	Variety	Native Location	Variety	Specimen	Genome
727.162.101.020.1	*amplissima*	W. Indies	Original	Burro Blanco	Caper
727.162.101.020.2	*portoricensis*	W. Indies (Puerto Rico)	Puerto Rico	Burro Blanco	Caper

727.162.101.03 **SPECIES 03** *baducca* Church Blossom Caper

Registration #	Variety	Native Location	Variety	Specimen	Genome
727.162.101.030.1	*baducca*	W. Indies	Original	Church Blossom	Caper
727.162.101.030.2	*cynophallophora*	W. Indies (Jamaica)	Jamaican	Church Blossom	Caper
727.162.101.030.3	*frondosa*	W. Indies	Frond	Church Blossom	Caper

727.162.101.04 **SPECIES 04** *corymbosa* Corymb Caper

727.162.101.05 **SPECIES 05** *cynophallophora* Jamaican

727.162.101.06 **SPECIES 06** *flexuosa* Limber Caper

727.162.101.07 **SPECIES 07** *hastata* Broad Leaf Caper

Registration #	Variety	Native Location	Variety	Specimen	Genome
727.162.101.070.1	*hastata*	W. Indies	Original	Broad Leaf	Caper
727.162.101.070.2	*coccolobifolia*	W. Indies	Pigeon Grape Leaf	Broad Leaf	Caper

727.162.101.08 **SPECIES 08** *incana* Hoary Leaf Caper

727.162.101.09 **SPECIES 09** *indica* Linguam Leaf Caper

727.162.101.10 **SPECIES 10** *mitchellii* Bumbil Orange Caper

727.162.101.11 **SPECIES 11** *sandwichiana* Hawaiian Caper

Registration #	Variety	Native Location	Variety	Specimen	Genome
727.162.101.110.1	*sandwichiana*	Pac. Is. (Hawaiian Is.)	Original	Hawaiian	Caper
727.162.101.110.2	*zoharyi*	Pac. Is. (Hawaiian Is.)	Zohary	Hawaiian	Caper

727.162.101.12 **SPECIES 12** *spinosa* Ribbed Caper

Registration #	Variety	Native Location	Variety	Specimen	Genome
727.162.101.120.1	*spinosa*	N. Africa, S. Europe, S. Asia, N. Australia	Original	Ribbed	Caper
727.162.101.120.2	*rupestris*	N. Africa, S. Europe, S. Asia, N. Australia	Rock Dwelling	Ribbed	Caper

727.162.101.13 **SPECIES 13** *tenuisiliqua* Tenuous Silique Caper

MORINGO SEED (*Moringa*)

Botanica ISBN:3-8331-1253-0, page 583, states that "The fruits are pod-like capsules". Botanica ISBN:3-8331-1253-0, page 583, states that for Moringa oleifera, "Fluted bright green 18 in fruit. The seeds and asparagus flavored pods are much used in Indian curries and the roots were once used for horseradish". Fruit, Berry and Nut Inventory Fourth Edition ISBN:978-1-882424-62-7, page 371, under Horseradish Tree, states that "Seeds are roasted and eaten".

The seeds, considered a fruit by Seed Saver's Exchange, are more properly classified as a vegetable.

Shrubs & Trees ISBN:0-7858-1789-1, on page 483, states that for *Moringa oleifera*; "The fruits are brown, pendulous, triangular pods up to 24 in long, splitting lengthwise into 3 when dry and containg abut 20 dark brown, papery-winged seeds. The fruits ripen in autumn. Native to the Arabian Peninsula and India, this species is now cultivated in many tropical countries for the food value of its leaves, flowers, pods, and roots".

GENUS 02 *Moringa* Moringo Pod

727.162.102.01 **SPECIES 01** *oleifera* Drumstick Moringo Pod

Registration #	Variety	Native Location	Variety	Specimen	Genome
727.162.102.010.1	*oleifera*	S. Asia (India)	Original	Drumstick	Moringo Pod
727.162.102.010.2	*pterygosperma*	S. Asia (India)	Wing Seed	Drumstick	Moringo Pod

727.162.102.02 **SPECIES 02** *peregrina* Peregrine Moringo Pod

Willow Order (*Salicales*)

Order 3 *Salicales* Willow

The National Plant Data Center, National Resources Conservation Service (NRCS), US Dept. of Agriculture (USDA), Baton Rouge, LA 70874-4490 USA. http://plants.usda.gov, database, ver. 4.0.4 creates the Order *Salicales* with **one** Family *Salicaceae*. In this book, in order to **limit** the number of families under Order *Violales* to **nine**, *Violales* Families are required to contain two or more *Genera*. The following Families have been moved from Order *Violales* to previously unused Order ***Salicales*** by the author of this book; *Datisaceae, Frankeniaceae, Passifloraceae, Stachyuraceae, Tamaricaceae*. These are the Families that bear the least resemblence to Family *Violaceae*, and are more properly placed in their own Order ***Salicales***

Passionflower Family (*Passifloraceae*)

Family 4 *Passifloraceae* Passion Flower

The Complete Book of Fruits, ISBN:0670-86752-7, page, 136, states that "These vary in size from that of a cherry to a coconut, and in color, coming in almost any shade from yellow to black. They are usually thick skinned with a juicy, acid, fragrant, sweet pulp inside, almost inseparable from smooth black seeds".

PASSION FRUIT (*Passiflora*)

GENUS 01 *Passiflora* — Passion Flower

727.163.401.01 **SPECIES 01** *actina* — Ray Yellow Passion Fruit Passion Flower
727.163.401.02 **SPECIES 02** *affinis* — Bracted Passion Flower
727.163.401.03 **SPECIES 03** *africana* — South African Passion Flower
727.163.401.04 **SPECIES 04** *alata* — Wing Stem Passion Flower

Registration #	Variety	Native Location	Variety	Specimen	Genome
727.162.401.040.1	*alata*	S. America (Amazon, Brazil, Peru)	Original	Wing Stem	Passion Flower
727.162.401.040.2	*rubra-glovia*	S. Asia (India)	Ruby Glow	Wing Stem	Passion Flower
727.162.401.040.3	*shannonii*	S. Asia (India)	Shannon	Wing Stem	Passion Flower

727.163.401.05 **SPECIES 05** *x allardi* — Allard Passion Flower
727.163.401.06 **SPECIES 06** *amethystina* — Amethyst Passion Flower
727.163.401.07 **SPECIES 07** *anadenia* — Glandless Passion Flower
727.163.401.08 **SPECIES 08** *antioquiensis* — Red Banana Passion Flower

Registration #	Variety	Native Location	Variety	Specimen	Genome
727.162.401.080.1	*antioquiensis*	S. America (Colombia)	Original	Red Banana	Passion Flower
727.162.401.080.2	*van-volxemii*	S. America (Colombia)	Van Volxem	Red Banana	Passion Flower

727.163.401.09 **SPECIES 09** *aurantia* — Orange Red Passion Flower
727.163.401.10 **SPECIES 10** *belotii (alata x caerulea)* — Belot Passion Flower
727.163.401.11 **SPECIES 11** *berteriana* — Berter Passion Flower
727.163.401.12 **SPECIES 12** *bicornis* — Wing Leaf Passion Flower

Registration #	Variety	Native Location	Variety	Specimen	Genome
727.162.401.120.1	*bicornis*	W. Indies	Original	Wing Leaf	Passion Flower
727.162.401.120.2	*pulchella*	W. Indies	Pretty	Wing Leaf	Passion Flower

727.163.401.13 **SPECIES 13** *biflora* — Two Flower Passion Flower
727.163.401.14 **SPECIES 14** *bilobata* — Two Lobe Passion Flower
727.163.401.15 **SPECIES 15** *bryonioides* — Cupped Passion Flower
727.163.401.16 **SPECIES 16** *caerulea* — Blue Passion Flower

Registration #	Variety	Native Location	Variety	Specimen	Genome
727.162.401.160.1	*caerulea*	S. America (Argentina, S.Brazil)	Original	Blue	Passion Flower
727.162.401.160.2	*amethystina*	S. America (Argentina, S.Brazil)	Amethyst	Blue	Passion Flower
727.162.401.160.3	*bluebirdii*	S. America (Argentina, S.Brazil)	Bluebird	Blue	Passion Flower
727.162.401.160.4	*constance-elliotii*	S. America (Argentina, S.Brazil)	Constance Elliot	Blue	Passion Flower
727.162.401.160.5	*eynsford-gemii*	S. America (Argentina, S.Brazil)	Eynsford Gem	Blue	Passion Flower
727.162.401.160.6	*grandiflora*	S. America (Argentina, S.Brazil)	Grand Flower	Blue	Passion Flower
727.162.401.160.7	*racemosa*	S. America (Argentina, S.Brazil)	Raceme	Blue	Passion Flower

727.163.401.17 **SPECIES 17** *capsularis* — White Star Passion Flower
727.163.401.18 **SPECIES 18** *ciliata* — Fringed Passion Flower

Registration #	Variety	Native Location	Variety	Specimen	Genome
727.162.401.180.1	*ciliata*	W. Indies	Original	Fringed	Passion Flower
727.162.401.180.2	*riparia*	W. Indies	Riverbank	Fringed	Passion Flower

727.163.401.19 **SPECIES 19** *cincinnata* — Crato Passion Flower
727.163.401.20 **SPECIES 20** *citrina* — Citrine Passion Flower
727.163.401.21 **SPECIES 21** *coccinea* — Scarlet Red Passion Flower
727.163.401.22 **SPECIES 22** *colvillii* — Colville Passion Flower
727.163.401.23 **SPECIES 23** *coriacea* — Bat Leaf Passion Flower
727.163.401.24 **SPECIES 24** *decaisnea* — Decaisne Passion Flower

Registration #	Variety	Native Location	Variety	Specimen	Genome
727.162.401.240.1	*decaisnea*	.	Original	Decaisne	Passion Flower
727.162.401.240.2	*fragrans*	.	Fragrant	Decaisne	Passion Flower

GRANADILLA (*Passiflora edulis*)

727.163.401.25 **SPECIES 25** *edulis* — Granadilla Passion Flower

Registration #	Variety	Native Location	Variety	Specimen	Genome
727.163.401.250.1	*edulis*	S. America (Brazil)	Original	Granadilla	Passion Flower
727.163.401.250.2	*bountiful-beautyi*	S. America (Brazil)	Bountiful Beauty	Granadilla	Passion Flower
727.163.401.250.3	*edgehillii*	S. America (Brazil)	Edgehill	Granadilla	Passion Flower
727.163.401.250.4	*flavicarpa*	S. America (Brazil)	Yellow	Granadilla	Passion Flower

Registration #	Variety	Native Location	Variety	Specimen	Genome
727.163.401.250.5	*flavicarpa-lilikoi*	S. America (Brazil)	Liliko Yellow	Granadilla	Passion Flower
727.163.401.250.6	*flavicarpa-mccainii*	S. America (Brazil)	McCain Yellow	Granadilla	Passion Flower
727.163.401.250.7	*flavicarpa-rotunda*	S. America (Brazil)	Round Yellow	Granadilla	Passion Flower
727.163.401.250.8	*frederickii*	S. America (Brazil)	Frederick	Granadilla	Passion Flower
727.163.401.250.9	*kahunai*	S. America (Brazil)	Kahuna	Granadilla	Passion Flower
727.163.401.251.0	*laceyi*	S. America (Brazil)	Lacey	Granadilla	Passion Flower
727.163.401.251.1	*panama-redii*	C. America (Panama)	Panama Red	Granadilla	Passion Flower
727.163.401.251.2	*purpurea*	S. America (Brazil)	Purple	Granadilla	Passion Flower
727.163.401.251.3	*purpurea-black-nightii*	S. America (Brazil)	Black Knight Purple	Granadilla	Passion Flower
727.163.401.251.4	*purpurea-goldii*	S. America (Brazil)	Gold Purple	Granadilla	Passion Flower
727.163.401.251.5	*purpurea-lilikoi*	S. America (Brazil)	Liliko Purple	Granadilla	Passion Flower
727.163.401.251.6	*purpurea-possum*	S. America (Brazil)	Possum Purple	Granadilla	Passion Flower
727.163.401.251.7	*red-roveri*	S. America (Brazil)	Red Rover	Granadilla	Passion Flower
727.163.401.251.8	*red-sunsetii*	S. America (Brazil)	Red Sunset	Granadilla	Passion Flower
727.163.401.251.9	*sweet-sunrisei*	S. America (Brazil)	Sweet Sunrise	Granadilla	Passion Flower

727.163.401.26 **SPECIES 26** *x exoniensis* Exon Passion Flower
727.163.401.27 **SPECIES 27** *filipes* Slender Passion Flower
727.163.401.28 **SPECIES 28** *foetida* Scarlet Passion Fruit Passion Flower

Registration #	Variety	Native Location	Variety	Specimen	Genome
727.163.401.280.1	*foetida*	S. USA, Mexico, W. Indies	Original	Scarlet Passion Fruit	Passion Flower
727.163.401.280.2	*arizonica*	S. USA (Arizona)	Original	Scarlet Passion Fruit	Passion Flower
727.163.401.280.3	*gossypiifolia*	S. USA (Texas), Mexico	Original	Scarlet Passion Fruit	Passion Flower
727.163.401.280.4	*hispida*	W. Indies	Original	Scarlet Passion Fruit	Passion Flower
727.163.401.280.5	*isthmia*	S. USA, Mexico	Original	Scarlet Passion Fruit	Passion Flower
727.163.401.280.6	*lanuginosa*	S. USA, Mexico	Original	Scarlet Passion Fruit	Passion Flower

727.163.401.29 **SPECIES 29** *gracilis* Crinkled Passion Flower
727.163.401.30 **SPECIES 30** *herbertiana* Herbert Passion Flower
727.163.401.31 **SPECIES 31** *incarnata* Maypop Passion Flower

Registration #	Variety	Native Location	Variety	Specimen	Genome
727.163.401.310.1	*incarnata*	SE. USA, Mexico	Original	Maypop	Passion Flower
727.163.401.310.2	*alba*	SE. USA, Mexico	White	Maypop	Passion Flower

727.163.401.32 **SPECIES 32** *kuangxii* Hua Shan Go Passion Flower
727.163.401.33 **SPECIES 33** *laurifolia* Laurel Leaf Passion Flower

Registration #	Variety	Native Location	Variety	Specimen	Genome
727.163.401.330.1	*laurifolia*	W. Indies (Jamaica), S. America (Tropical)	Original	Laurel Leaf	Passion Flower
727.163.401.330.1	*golden-bell-applei*	W. Indies (Jamaica), S. America (Tropical)	Golden Bell Apple	Laurel Leaf	Passion Flower
727.163.401.330.1	*water-lemonii*	W. Indies (Jamaica), S. America (Tropical)	Water Lemon	Laurel Leaf	Passion Flower

727.163.401.34 **SPECIES 34** *ligularis* Sweet Granadilla Passion Flower
727.163.401.35 **SPECIES 35** *lutea* Yellow Passion Flower

Registration #	Variety	Native Location	Variety	Specimen	Genome
727.163.401.350.1	*lutea*	SE. USA, Mexico	Original	Yellow	Passion Flower
727.163.401.350.2	*glabriflora*	SE. USA, Mexico	Bald Flower	Yellow	Passion Flower

727.163.401.36 **SPECIES 36** *maliformis* Sweetcup Conch Apple Passion Flower
727.163.401.37 **SPECIES 37** *manicata* Long Sleeve Red Passion Flower
727.163.401.38 **SPECIES 38** *mexicana* Mexican Passion Flower
727.163.401.39 **SPECIES 39** *mixta* Mixed Banana Passion Flower
727.163.401.40 **SPECIES 40** *mollissima* Banana Passion Flower

Registration #	Variety	Native Location	Variety	Specimen	Genome
727.163.401.400.1	*mollissima*	S. America (Andes Mtns.)	Original	Banana	Passion Flower
727.163.401.400.2	*pink-bananai*	S. America (Andes Mtns.)	Pink	Banana	Passion Flower

727.163.401.41 **SPECIES 41** *morifolia* Woodland Passion Flower

Registration #	Variety	Native Location	Variety	Specimen	Genome
727.163.401.410.1	*morifolia*	S. USA, Mexico	Original	Woodland	Passion Flower
727.163.401.410.2	*warmingii*	S. USA, Mexico	Warming	Woodland	Passion Flower

727.163.401.42 **SPECIES 42** *multiflora* Many White Passion Flower
727.163.401.43 **SPECIES 43** *murucuja* Virgin Island Passion Flower
727.163.401.44 **SPECIES 44** *pallens* Pineland Passion Flower
727.163.401.45 **SPECIES 45** *pfordtii* Tiger Passion Flower
727.163.401.46 **SPECIES 46** *platyloba* Flat Lobe Grape Passion Flower
727.163.401.47 **SPECIES 47** *quadrangularis* Four Angled Passion Flower

Registration #	Variety	Native Location	Variety	Specimen	Genome
727.163.401.470.1	*quadrangularis*	C. & S. America (Tropical), W. Indies	Original	Four Angled	Passion Flower
727.163.401.470.2	*gigantea*	C. & S. America (Tropical), W. Indies	Giant Granadilla	Four Angled	Passion Flower
727.163.401.470.3	*golden-applei*	C. & S. America (Tropical), W. Indies	Golden Apple	Four Angled	Passion Flower

727.163.401.48 **SPECIES 48** *racemosa* Raceme Red Passion Flower
727.163.401.49 **SPECIES 49** *reflexiflora* Long Tube Magenta Passion Flower
727.163.401.50 **SPECIES 50** *rubra* Duchman's Red Passion Flower
727.163.401.51 **SPECIES 51** *serratifolia* Red Purple Passion Fruit Passion Flower
727.163.401.52 **SPECIES 52** *serratodigitata* Tagua Tagua Passion Flower
727.163.401.53 **SPECIES 53** *sexflora* Goat's Foot Passion Flower
727.163.401.54 **SPECIES 54** *srilankensis* Sri Lanka Passion Flower
727.163.401.55 **SPECIES 55** *suberosa* Corky Stem Passion Flower

Registration #	Variety	Native Location	Variety	Specimen	Genome
727.163.401.550.1	*suberosa*	W. Indies	Original	Corky Stem	Passion Flower
727.163.401.550.2	*diabolica-cucurbita*	W. Indies	Devil's Pumkin	Corky Stem	Passion Flower
727.163.401.550.3	*hue-hue-haolei*	W. Indies	Hue Hue Haole	Corky Stem	Passion Flower
727.163.401.550.4	*indigofera*	W. Indies	Indigo Berry	Corky Stem	Passion Flower
727.163.401.550.5	*morita*	W. Indies	Morita	Corky Stem	Passion Flower
727.163.401.550.6	*pallida*	W. Indies	Pale	Corky Stem	Passion Flower

727.163.401.56 **SPECIES 56** *subpeltata* Grenadine Passion Flower
727.163.401.57 **SPECIES 57** *tenuiloba* Bird Wing Passion Flower
727.163.401.58 **SPECIES 58** *tuberosa* Tuber Passion Flower
727.163.401.59 **SPECIES 59** *tulae* Mountain Love In The Mist Passion Flower
727.163.401.60 **SPECIES 60** *x violacea* Palm Leaf Passion Flower
727.163.401.61 **SPECIES 61** *vitifolia* Perfumed Passion Flower
727.163.401.62 **SPECIES 62** *vitifolia x coccinea* Scarlet Perfumed Passion Flower

Registration #	Variety	Native Location	Variety	Specimen	Genome
727.163.401.620.1	*coccinea*	C. Amer.(Nicaragua), S.Amer. (Peru, Venezuela)	Original	Scarlet Perfumed	Passion Flower
727.163.401.620.2	*flamea*	C. Amer.(Nicaragua), S.Amer. (Peru, Venezuela)	Scarlet Flame	Scarlet Perfumed	Passion Flower

INDEX

Fruit

CD File Format

(Microsoft Excel Spreadsheet .xls)

A	B	C	D	E	F	G	H	I	J
Regist. #	**Genus**	**Species**	**Infraspecies**	**Author, Year**	**Source**	**Location**	**Variety**	**Specimen**	**Genome**

K	L	M	N	O	P	Q	R	S	T
Behavior	**Habit**	**Cross Ref.**	**Resource**	**Domain**	**Kingdom**	**Subkingdom**	**Phylum**	**Class**	**Order**

U	V	W	X	Y	Z
Family	**Status**	**Habitat**	**Temp. Zone**	**Height**	**Length**

AA	AB	AC	AD	AE	AF	AG	AH	AI
Silhouette	**Feature 1**	**F 1 Color**	**Feature 2**	**F 2 Color**	**Feature 3**	**F 3 Color**	**Feature 4**	**F 4 Color**

A Registration #	taxonomic number in format: **ksp.cof.ggs.svv**, **k** = kingdom #, **s** = subkingdom #, **p** = phylum #, **c** = class #, **o** = order #, **f** = family #, **gg** = genus #, **s.s** = species #, **vv** = variety #	
B Genus	general name	**Latin**
C Species	specific name	**Latin**
D Infraspecies	subspecies or variety name	**Latin**
E **Author, Year**	scientist who first published species or variety, or book abbrev. and year of publication	**English**
F **Source**	technical journal or book-page #-species/var. # or mail order company name abbrev.	**English**
G **Location**	indigenous (native) location by Subcontinent (Country) or by Country (City, State)	**English**
H **Variety**	infra-species **UNIQUE OFFICIAL VALID ENGLISH NAME**	**English**
I **Specimen**	species **UNIQUE OFFICIAL VALID ENGLISH NAME**	**English**
J **Genome**	genus - RNA/DNA genetic group **UNIQUE OFFICIAL VALID ENGLISH NAME**	**English**
K **Behavior**	**microbe:** Antibiotic, Parasite, Pathogen, **plant:** herb, vegetable, fruit, flower, foliage, succulent, **animal:** Parasite, Pathogen, Predator, Symbiote	**English**
L **Habit**	**microbe:** Free Living, Symbiote, Parasite, Pathogen, **plant:** Annual, Biennial, Perennial, Trailing or Climbing Vine, Bush, Shrub, Tree, **animal:** Herbivore, Carnivore, Insectivore, Omnivore, Scavenger	**English**
M **Cross Ref.**	dataset abbreviation-sequence number for Species or Infraspecies	**English**
N **Resource**	web site address or Intl. Std. Book Number (ISBN) for Species and Infraspecies	**English**
O Domain	*Protista, Plantae or Animalia*	**Latin**
P Kingdom	kingdom name: *Materia, Microbeavirus, Plantaevirus, Animaliavirus, Microbea, Sporophyta,* ***Spermatophyta*** **(Seed Plant)**, *Achordata,* **or** *Chordata*	**Latin**
Q Subkingdom	Subkingdom name for one of 38 Subkingdoms	**Latin**
R Phylum	Phylum name	**Latin**
S Class	Class name	**Latin**
T Order	Order name	**Latin**
U Family	Family name	**Latin**
V **Status**	population: **A**bundant, **B**elow optimum, **C**onserved, **D**ecimated, **E**ndangered, **F**ailed, **G**enocide	**English**
W **Habitat**	1. ecosystem zone #, 2. ecotype - desert, wetland, forest, inland plain, coastal plain	**English**
X **Temp. Zone**	1. minimum temperature (hardiness) zone #, 2. maximum temperature (heat) zone #	**English**

	Nonliving Matter	**Microbe**	**Plant**	**Animal**	
Y **Height**	cleavage	colony width micrometers	min-max ft. or in.	min-max ft. or in.	**English**
Z **Length**	fracture	colony length micrometers	min-max spread	min-max ft. or in.	**English**
AA **Silhouette**	crystal habit	colony shape	**plant** outline type	number of legs	**English**
AB **Feature 1**	chemical formula	colony cell count	**leaf** shape & length	leg length	**English**
AC **F 1 Color**	appearance color	colony color	**leaf** color	fur/skin color	**English**
AD **Feature 2**	hardness mohs 1-10	cell shape & length	**flower** arrangement, shape & length	ear/horn shape & length	**English**
AE **F 2 Color**	strike/streak color	colony color	**flower** color	ear/horn color	**English**
AF **Feature 3**	lattice structure	spore type & length	**fruit** type & length	foot or paw or hoof	**English**
AG **F 3 Color**	pigment color	spore color	**fruit** color	foot/paw/hoof color	**English**
AH **Feature 4**	specific gravity	cell wall thickness	**bark** texture	eye size & shape	**English**
AI **F 4 Color**	Luster	cell wall color	bark color	eye color	**English**

Spreadsheet Column W, Item 1 – Ecosystem Zone

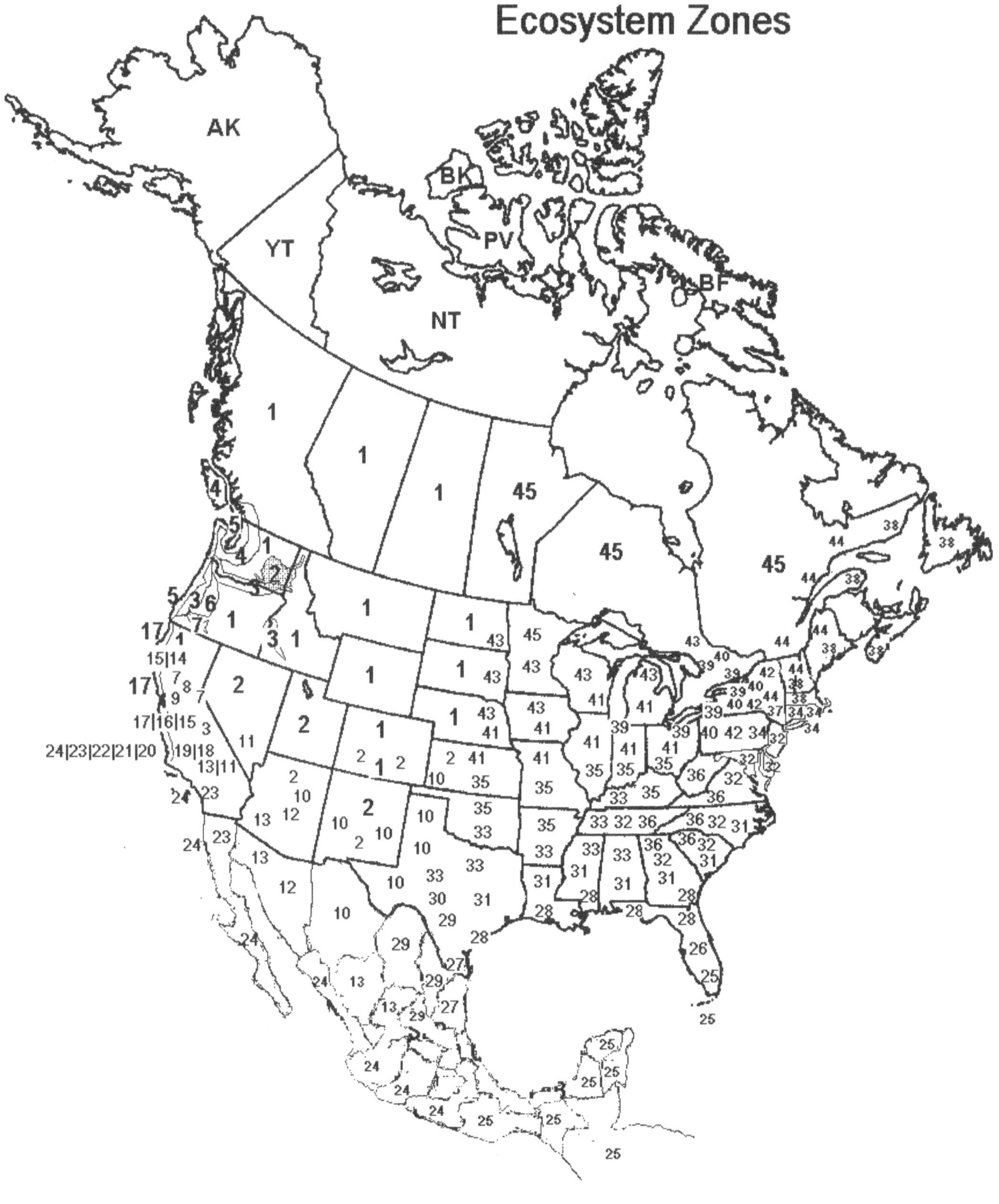

Spreadsheet Column W, Item 1 - Ecosystem Zone

#	Ecosystem Zone Name	Growing [Primary] Season
01 Valley:	NW. Extreme Cold Winter	May[Jun-Aug]Sep
02 Valley:	W. Very Cold Winter	Apr[May-Sep]Oct
03 Hillside:	Mild High Elevation/Interior	May[May-Sep]Oct
04 Valley:	NW. Cold Winter	Apr[May-Sep]Oct
05 Coastal:	NW. Coast / Puget Sound	Feb,Mar,Apr[Apr-Oct]Nov
06 Valley:	Williamette Valley	Feb,Mar[Apr-Oct]Nov
07 Valley:	Digger Pine Belt/Rogue River V.	Apr[May-Sep]Oct
08 Valley:	CA Central Valley-Warm	Feb[Mar-Nov]Dec
09 Valley:	CA CentralValley-Hot	Jan[Feb-Nov]Dec
10 Desert:	W. High Desert-3300ft	Mar[Apr-Oct]Nov
11 Desert:	W. Elevated Desert-2500feet	Mar[Apr-Oct]Oct
12 Desert:	W. IntermediateDesert-1500ft	Mar[Apr-Oct]Nov
13 Desert:	W. Low Desert-500ft	Feb[Mar-May]Jun
13 Desert:	W. Low Desert-500ft	Aug[Sep-Nov]Dec
14 Coastal:	W.Coast Inland-Marine	Jan[Feb-Nov]Dec
15 Coastal:	W. Coast Range-Cool Winter	Jan,Feb[Mar-Nov]Dec
16 Coastal:	W. Coast Inland-ThermalBelt	Jan[Feb-Nov]Dec
17 Coastal:	W. Coast-Northern Marine	Mar[Apr-Oct]Nov
18 Valley:	CA Interior Valley	Mar[Apr-Nov]Dec
19 Valley:	CA Int. Valley-Thermal Belt	Jan,Feb[Mar-Nov]Dec
20 Valley:	CA Occas. Ocean Influence	Jan,Feb[Mar-Nov]Dec
21 Valley:	CA Occ. Oc. Infl-Thermal Belt	Jan,Feb[Mar-Nov]Dec
22 Plains:	Pac. Coast Inland-Cold Winter	Jan,Feb[Mar-Nov]Dec
23 Coastal:	Pacific Coast-Thermal Belt	Jan[Jan-Dec]Dec
24 Coastal:	PacificCoast-SouthernMarine	Jan[Jan-Dec]Dec
25 Tropical		Jan[Jan-Dec]Dec
26 Plains:	Central Florida Subtropical	Jan[Feb-Nov]Dec
27 Valley:	Lower Rio Grande Valley	Jan,Feb[Mar-Nov]Dec
28 Plains:	Gulf Coast Plains	Jan,Feb[Mar-Nov]Dec
29 Plains:	Atlantic Interior Plains	Mar[Apr-Nov]Dec
30 Hillside:	CentralTexasHillCountry	Mar[Apr-Nov]Dec
31 Plains:	Southern Interior Plains	Mar[Apr-Oct]Nov
32 Hillside:	Atlantic Foothill Plains	Mar[Apr-Oct]Nov
33 Hillside:	Atlantic Foothills	Mar[Apr-Oct]Nov
34 Valley:	Atlantic Mountain Valley	Apr[May-Oct]Nov
35 Mountain:	Central Mountains	Apr[May-Oct]Nov
36 Mountain:	Appalachian Mountains	Apr[May-Oct]Nov
37 Valley:	Hudson V. / Appalachian Plateau	Apr[May-Sep]Oct
38 Plains:	New England Int. / Canada Coast	Apr[May-Sep]Oct
39 Coastal:	Great Lakes Shoreline	Apr[May-Sep]Oct
40 Plains:	E. Great Lakes Inland Plains	May[Jun-Aug]Sep
41 Plains:	W. Great Lakes Inland Plains	May[Jun-Sep]Oct
42 Valley:	Saint Lawrence Valley & Mtns.	May[Jun-Sep]Oct
43 Valley:	Upper Mississippi Valley	May[Jun-Sep]Oct
44 Mountain:	New England / E.Canada Mtns.	May[Jun-Aug]Sep
45 Plains:	NE. Extreme Cold Winter	Jun[Jul-Aug]Sep

Spreadsheet Column W, Item 2 – Ecotype Name

Ecotype Name USA States
01 Forest: AtlanticBorealForest ME,MI,NH,NY,VT,WI
02 Forest: AtlanticBogForest CT,MA,ME,NH,NJ,NY,OH,PA,VT
03 Forest: AtlanticJackPineForest MI,MN,WI
04 Forest: AtlanticFir-Spruce-BirchForest ME,NH,VT
05 Forest: AtlanticDeciduousHardwoodForest ME,MI,MN,NH,NY,VT
06 Forest: AtlanticBeech-MapleForest IL,MI,NY,OH
07 Forest: AtlanticMaple-BasswoodForest MN,WI
08 Forest: AtlanticOak-HickoryForest MA,MD,NJ,NY,OH,PA,VA
09 Forest: AtlanticRiver-FloodPlainForest CT,DE,MA,MD,ME,NH,NJ,VA
10 Forest: AtlanticPine-OakForest CT,DE,MA,MD,NJ,VA
11 Forest: AtlanticLoblolly-VirginiaPineForest DE,MD,NJ,VA
12 Forest: AtlanticMountainBasswoodForest KY,MD,OH,VA,WV
13 Forest: AtlanticSavannaOakForest MN,WI
14 Forest: AtlanticOak-GumForest MD
15 Field: AtlanticPlainsOpenField NEUS
16 Field: AtlanticMountainOpenField NEUS
17 Forest: AtlanticSwampMapleForest CT,DE,MA,MD,ME,NH,NJ
18 Forest: AtlanticSwampWhiteCedarForest CT,DE,ME,MD,NH,NJ,VA
19 Forest: AtlanticSwampBaldCypressForest DE,MD,VA
20 Forest: AtlanticSwampMixedHardwoodForest DE,MD,VA
21 Forest: AtlanticSwampShrubForest NEUS
22 Wetland:AtlanticWetlandSaltMarsh NEUS
23 Wetland: AtlanticWetlandFreshwaterMarsh NEUS
24 Wetland: AtlanticWetlandOceanBeachGrass NEUS
25 Wetland: AtlanticWetlandFreshwaterShoreGrass NEUS
26 Wetland: AtlanticWetlandFloodplainField NEUS
27 Wetland: AtlanticWetlandPondWeed NEUS
28 Wetland: AtlanticWetlandRiverWeed NEUS
29 Forest: AtlanticNatureTrails MD
30 Forest: CentralPrairieRiparianForest IL,IA,MO,AR,OK,KS,NB,ND,SD
31 Forest: CentralBlackHillsForest SD,WY
32 Forest: CentralGulfCoastLiveOakForest TX
33 Forest: CentralEdwardsPlateauForest TX
34 Forest: CentralLowerRioGrandeForest TX,MX
35 Forest: CentralEastTexasPineForest TX
36 Forest: RockyMountainGreatBasinDesert NWUS,SWUS
37 Forest: RockyMountainShortGrassPrairie MT,WY,CO,NM
38 Forest: RockyMountainPinyon-JuniperForest UT,CO,NM,AZ
39 Forest: RockyMountainPonderosaPineForest ND,SD,MT,WY,NV,CO
40 Forest: RockyMountainAspenGroveForest NWUS,SWUS
41 Forest: RockyMountainLodgepolePineForest NWUS,SWUS
42 Forest: RockyMountainRiparianForest NWUS,SWUS
43 Forest: RockyMountainSpruce-FirForest NWUS,SWUS
44 Forest: RockyMountainHighPineForest NWUS,SWUS
45 Forest: RockyMountainTimberlineAlpineTundra NWUS,SWUS
46 Desert: CentralArroyoAndDesertScrub SCUS,SWUS,MX
47 Forest: CentralMadreanFoothillForest SCUS,MX
48 Forest: PacificMadreanFoothillForest AZ
49 Forest: PacificCanyonRiparianForest AZ
50 Forest: PacificMountainForest AZ,NM
51 Desert: PacificChihuahuaDesert TX,NM,AZ
52 Forest: PacificGiantSaguroCactusForest AZ,MX,BCA
53 Field: GrasslandCoastalShoreline/BeachOcean
54 Field: GrasslandCoastalPlainsNearShoreOcean
55 Field: GrasslandCoastalPlainsInlandOcean
56 Field: GrasslandCoastalShoreline/BeachLake
57 Field: GrasslandCoastalPlainsNearShoreLake
58 Field: GrasslandCoastalPlainsInlandLake
59 Field: GrasslandHills
60 Field: GrasslandFoothills
61 Field: GrasslandFoothillPlains/Plateau
62 Field: GrasslandValleyRiver
63 Field: GrasslandValleyMountain
64 Field: GrasslandValleyInland
65 Field: GrasslandInteriorLowElevation500ft
66 Field: GrasslandInteriorMediumElevation1500ft
67 Field: GrasslandInteriorMediumHighElevation2500ft
68 Field: GrasslandInteriorHighElevation3000ft
69 Plains: AtlanticArcticTundra
70 Plains: PacificArcticTundra
71 Plains: AtlanticBorealTundra
72 Plains: PacificBorealTundra
73 Forest: AtlanticArcticConiferForest
74 Forest: PacificArcticConiferForest
75 Forest: AtlanticBorealConiferForest
76 Forest: PacificBorealConiferForest

Spreadsheet Column X, Item 1 – Minimum Temperature Zone

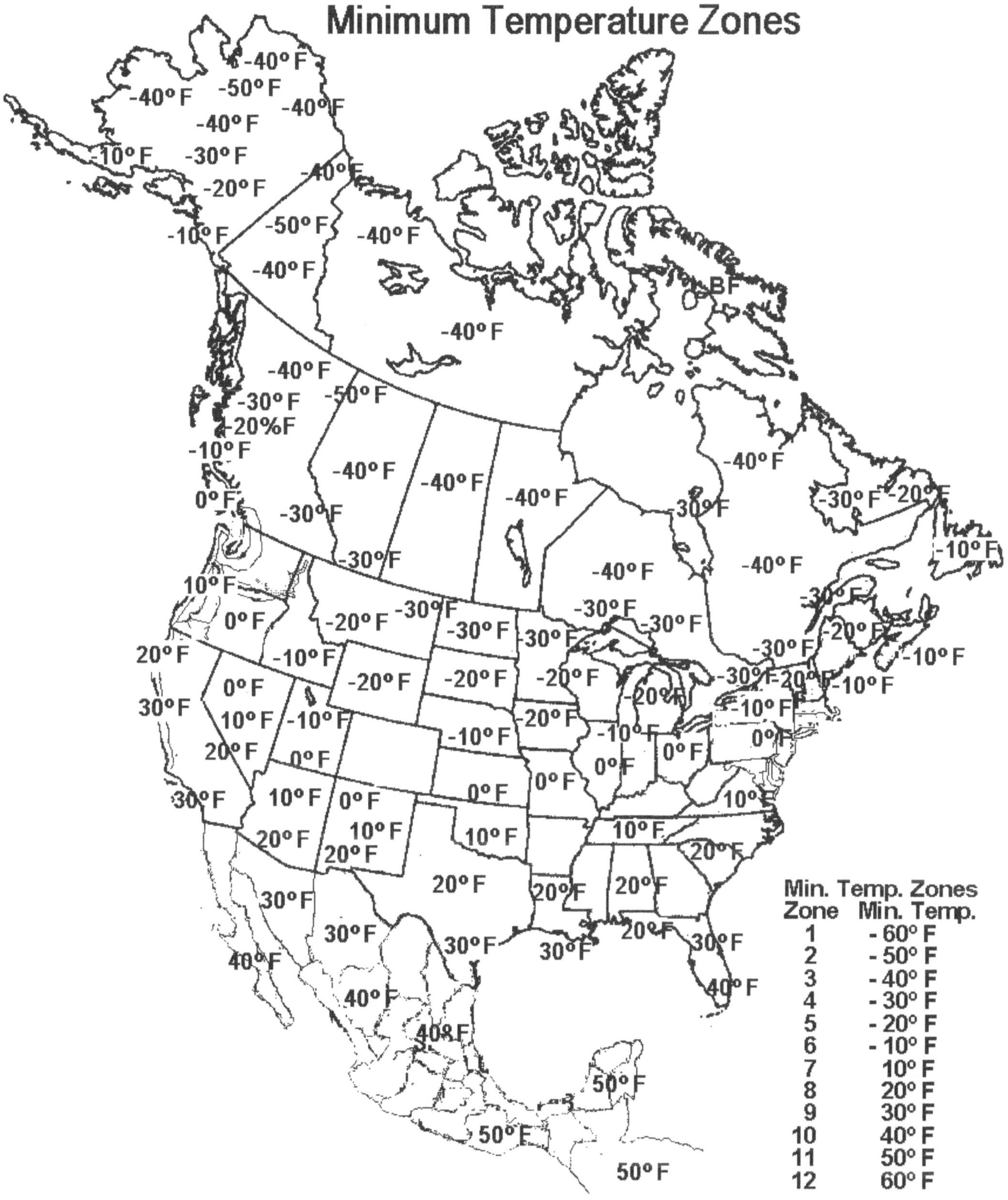

Spreadsheet Column X, Item 2 – Maximum Temperature Zone

Days Per Year Above 86°F

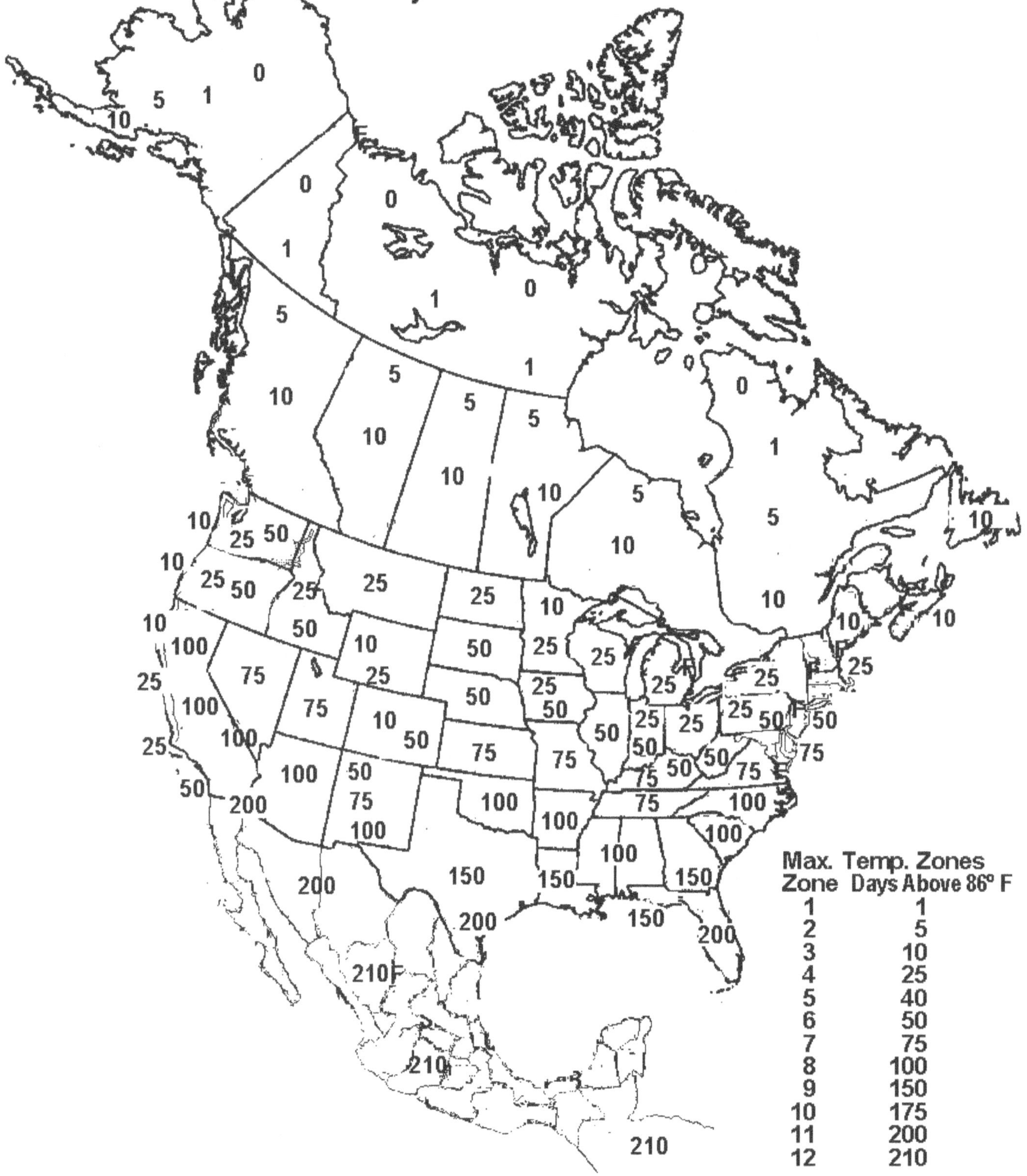

Max. Temp. Zones

Zone	Days Above 86° F
1	1
2	5
3	10
4	25
5	40
6	50
7	75
8	100
9	150
10	175
11	200
12	210

CD Order Form

Mail to: Lawrence J. Fusillo
280 Picadilly Sq. Apt. E-2
Athens, GA 30605

Item Number			
N01	Nonliving Matter Species CD	includes Elements, Minerals, Soils, Sediments & Rocks	$10.00
N02	Virus Species CD	includes all 5000 known microbe, plant & animal Viruses	$10.00
N03	Microbe Species CD	includes all Bacteria, Protophytes (Plankton) & Protozoans	$10.00
N04	Spore Plant Species CD	includes all Algae, Fungus, Moss & Ferns	$10.00
N05	Orchid Species CD	includes all known species and varieties of Orchids	$10.00
N06	Vegetable & Herb Variety CD	includes all known Vegetables & Herbs	$10.00
N07	Fruit Variety CD	includes all known edible Fruit Varieties	$10.00
N08	Flower Variety CD	includes florist quality Flower Plant Varieties	$10.00
N09	Lower Invertebrate CD	includes all Seanettle, Coral, Sponge, Snail & Worm Species	$10.00
N10	Shellfish Invertebrate CD	includes all Lampshells, Sea Urchins, Small Crustaceans (microshrimp, microcrayfish), Large Crustaceans (crabs, shrimp, lobsters, decapods, krill, squill, crayfish), Maxillopod Crustaceans (fish lice, barnacles, copepods, mystic crayfish), Bivalve Mollusks (arkshells, clams, mussels, oysters), Tentacle Shells (nautilus, argonaut), Capshells, Chitons, Tubeshells (tuskshells, toothshells), Gastropods (limpets, sea snails, ceriths, cowries, periwinkles, hornshells, conch shells, sea hares, sea butterflies)	$10.00
N11	Beetle Invertebrate CD	includes 144,324 of the 315,000 beetle species & varieties	$10.00
N12	Arthropod Invertebrate CD	includes **Chelicers** (mites, ticks, spiders, scorpions) & **Insects** (centipede, millipede, ant, bee, wasp, cockroach, mantis, webspinner, walkingstick, shorthorn bug, longhorn bug, louse, thrips, zorapter, fly, moth, butterfly, lacewing, dameselfly, dragonfly, stonefly, flea, caddis fly)	$10.00
N13	Fish Vertebrate CD	includes all known vertebrate fish	$10.00
N14	Higher Vertebrate CD	includes all known amphibians, reptiles, birds, mammals	$10.00
N15	FNCS Master CD	includes contents of CD N01 through CD N14	$100.00

Ship To: Name ______________________________

Address ______________________________

____

City ____________________ State______ Zip________

Quantity	Item Number	CD Title	Price

Georgia Resident 6.6% Sales Tax $ ________

Postage & Handling $3.50

Amount Enclosed $ ________

CHECK OR MONEY ORDER ONLY

Make Check Payable to Lawrence J. Fusillo

About the Author

Lawrence Fusillo was born in Greenbelt Maryland on May 22 1949. While in elementary school, he was introduced to plant life, by his neighbor, Anna Belle Owens, who had a large garden in her back yard. His interest in biology continued in junior high school in a biology club, and in high school when he was a volunteer in a microbiology laboratory at the Veteran's Administration hospital in Washington DC. He received a B. A. in Physics from John's Hopkins University in 1971, and an M. S. in Applied Physics from the University of New Orleans in 1981.

He became interested in the names of colors in 1996, working at the Naval Air Development Center in Warminster Pennsylvania while developing software to display the colors of seafloor sediment. He found that most colors were named after plants that exhibited the color in the form of flowers. He began studying the relationship between plants, animals, minerals, and color names and discovered that the Linnaeus *Genus species* Latin classification system for plants, animals and minerals had not been updated since 1758.

After a 26 year career designing magnetometer, sonar and radar software for the US Navy first as a government employee (1972-1981) and later as a US Navy contractor (1982-1998), he designed air traffic control software as a contractor for the Federal Aviation Administration (1999-2001). He began background research for writing scientific books in May of 2001.

The "Fusillo Nature Classification System Fruit Variety Checklist" with category names in both Latin & English, is designed to replace the Linnaeus Classification System and all subsequent classification schemes. The categories Domain, Kingdom, Subkingdom, Phylum, Class, Order, Family, Genus, species and variety were assigned both in English and in Latin. A companion CD "N07 Fusillo Nature Classification System Fruit Variety Checklist" contains a Microsoft Excel Spreadsheet (.xls) implementation of the Fruit Varieties which can be imported into database programs.

Books by Lawrence J. Fusillo

The Fusillo Nature Classification System (FNCS) containing all living organisms was published in;

	Fusillo Nature Classification System (FNCS) Title		Year
1	"Anna Belle Nature Trail Guide 1st Edition"	ISBN:0-7414-4201-9	Jul 2007
2	"Fusillo Nature Classification System Genus & Higher Taxa 1st Edition"	ISBN:0-7414-4495-3	Jan 2008
3	"FNCS Nonliving Matter Species Checklist 1st Edition"	ISBN:0-7414-4459-3	Mar 2008
4	"FNCS Virus Species Checklist 1st Edition"	ISBN:0-7414-5214-6	Mar 2009
5	"FNCS Microbe Species Checklist 1st Edition"	ISBN:0-7414-5475-0	Aug 2009
6	"FNCS Orchid Species Checklist 1st Edition"	ISBN:0-7414-5668-0	Dec 2009
7	"FNCS Spore Plant Species Checklist 1st Edition"	ISBN:0-7414-5878-0	May 2010
8	"FNCS Fruit Variety Checklist 1st Edition"	ISBN:0-7414-6441-1	Mar 2011
9	"FNCS Herb & Vegetable Variety Checklist 1st Edition"	future	
10	"FNCS Flower Variety Checklist 1st Edition"	future	
11	"FNCS Seed Plant Species Checklist 1st Edition"	future	
12	"FNCS Invertebrate Coral, Nettle, Sponge, & Worm Species Checklist 1st Edition"	future	
13	"FNCS Invertebrate Shellfish Species Checklist 1st Edition"	future	
14	"FNCS Invertebrate Beetle Species Checklist 1st Edition"	future	
15	"FNCS Invertebrate Spider & Insect Species Checklist 1st Edition"	future	
16	"FNCS Vertebrate Fish Species Checklist"	future	
17	"FNCS Vertebrate Amphibian, Reptile, Bird, & Mammal Species Checklist 1st Edition"	future	
1	Fusillo Knowledge Infrastructure with Dewey Decimal & CIP Cross Reference	future	
2	FKI PC Hardware Physics 1st Edition	future	
3	FKI PC Hardware & Operating System Advances 1st Edition	future	
4	FKI PC Hardware Components Catalog 1st Edition	future	
5	FKI PC Programmer's Guide 1st Edition	future	
6	FKI PC Color Language and Dictionary of Names 1st Edition	future	
7	FKI PC Programmer's Data Dictionary 1st Edition	future	
8	FKI PC User's Guide to Computer Books, Magazines & User Groups 1st Edition	future	
9	FKI PC User's Guide to Educational Software 1st Edition	future	
10	FKI PC User's Guide to Educational Web Sites 1st Edition	future	
11	FKI Body Tuneup Organic Foods, Air Filter, & Exercise 1st Edition	future	
12	FKI Mind Tuneup Brain Foods and Mental Agility Exercises 1st Edition	future	

Made in United States
Cleveland, OH
06 November 2026

47787690R00295